Genetically Modified Organisms

Edited by Huseyin Tombuloglu and Guzin Tombuloglu

Published in London, United Kingdom

Genetically Modified Organisms
http://dx.doi.org/10.5772/intechopen.111328
Edited by Huseyin Tombuloglu and Guzin Tombuloglu

Contributors
América Padilla-Viveros, Augustine Antwi-Boasiako, Augustine Boakye Boateng, Beatriz Xoconostle-Cázares, Cheng Chen, Deepa Beniwal, Gauri Nerkar, Harnoor Kaur Dhillon, Isaac Newton Boakye-Mensah, Ivy Odi Ahiamadia, Joseph Gyau, José Luis Millán, Kiran Jagtap, Koichi Miyake, Laura Claret Triana Vidal, Le Qu, Matilda Frimpong, Mukta Nagpurkar, Padmore Adu-Antwi, Radhika Velankar, Richard Adu Amoah, Roberto Ruiz-Medrano, Rosa Obando-González, Shivani Chauhan, Sonoko Narisawa, Tae Matsumoto, Tammy Walker-Smith, Xin Pan, Yaoming Li, Yoatzin Guadalupe Domínguez-Fernández

First published in London, United Kingdom, 2024 by IntechOpen
IntechOpen is the global imprint of INTECHOPEN LIMITED, registered in England and Wales, registration number: 11086078, 167-169 Great Portland Street, London, W1W 5PF, United Kingdom

British Library Cataloguing-in-Publication Data
A catalogue record for this book is available from the British Library

Additional hard and PDF copies can be obtained from orders@intechopen.com

Genetically Modified Organisms
Edited by Huseyin Tombuloglu and Guzin Tombuloglu
p. cm.

This title is part of the Genetics Book Series, Volume 3
Topic: Genomics
Series Editor: Kenji Ikehara
Topic Editor: Irina Vlasova-St. Louis

Print ISBN 978-0-85466-962-2
Online ISBN 978-0-85466-961-5
eBook (PDF) ISBN 978-0-85466-963-9
ISSN 3049-7094

For EU product safety concerns:
IN TECH d.o.o., Prolaz Marije Krucifikse Kozulić 3, 51000 Rijeka, Croatia,
info@intechopen.com or visit our website at intechopen.com.

IntechOpen Book Series

Genetics

Volume 3

Aims and Scope of the Series

"Genetics," which has been proud of its tradition since Mendel presented his research results in 1865, initially progressed quite slowly due to simple observational approaches of individuals and groups. However, the discovery of double-stranded DNA by Watson and Crick about 70 years ago triggered rapid progress in life sciences, including genetics, which was primarily conducted using Escherichia coli and bacteriophages infecting E. coli. Subsequently, genetics has achieved remarkable developments, such as understanding genetic disorders, including cancers, through research on the biogenesis and differentiation of plants and animals.
The two topics of this book series - Human Genetics, and Genomics - will address important areas of advancement in genetics.
Human Genetics: After fundamental genetics, initially studied with the main goal of revealing the functions of individual genes and proteins, genetics expanded from understanding the genetic system itself to understanding many infectious diseases caused by bacteria and viruses. Consequently, human beings are now overcoming infectious diseases by developing medicinal chemicals, including antibiotics and vaccines. However, genetic disorders remain challenging to cure up to now. Nevertheless, even the cure for them, including various cancers, is coming closer to reality due to the rapid progress of human genetics. In this way, the welfare of human life continues to improve, and even longevity, which was once a dream, has been achieved to some extent in recent years.
Genomics: On the other hand, the understanding of the comprehensive interrelationship of whole genes or whole proteins functioning in one organism has become possible now, as research has entered the era of genomics, owing to the rapid progress of base sequence analysis and bioinformatics. The development of genomics has further made it possible to understand the evolutionary processes of organisms through comparative studies among the genomes of many organisms.
This book series will discuss the findings obtained during the advancement of human genetics and genomics. It is also expected that this series will trigger the formation of a better world composed of human beings and all other organisms on Earth through discussions of research results obtained under the development of general genetics.

Meet the Series Editor

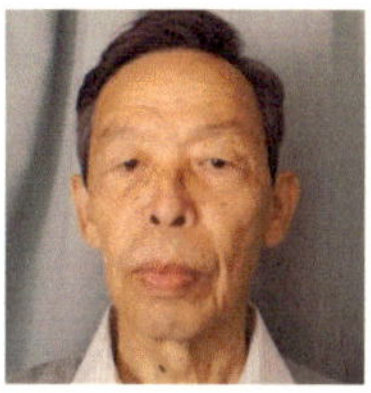

Kenji Ikehara graduated from the Department of Industrial Chemistry, Faculty of Engineering, Kyoto University in 1968. He received his B. Eng. (1968) and subsequently earned M. Eng. (1970) and D. Eng. (1976) degrees from Kyoto University. He began his career as a research associate in the Faculty of Science at the University of Tokyo before moving on to become an associate professor in the Faculty of Science at Nara Women's University. He was later promoted to professor and subsequently served as the dean of the Faculty of Science at Nara Women's University. Additionally, he held the position of director at the Nara Study Center of the Open University of Japan. For approximately 15 years, he focused his research on sporulation initiation of Bacillus subtilis. Later, he shifted his focus to the origins and evolutionary processes of microbial genes, the genetic code, proteins, and life. He has proposed several hypotheses, including the GC-NSF(a) hypothesis on the origin of genes, the GNC-SNS hypothesis on the genetic code, the protein 0th-order structure hypothesis on the origin of proteins, and the [GADV]-protein world hypothesis (GADV hypothesis) on the origin of life. Furthermore, he served as the local chair of the International Conference, Origin 2014, held in Nara in 2014.

Meet the Volume Editors

Huseyin Tombuloglu, Ph.D., is an associate professor at the Institute of Research and Medical Consultation (IRMC) of Imam Abdulrahman bin Faisal University, Dammam, Saudi Arabia. He received his BSc in Molecular Biology and Genetics in 2007, and studied at the University of Groningen, the Netherlands. He obtained an MSc in 2010 and a Ph.D. in Biotechnology in 2014. He became an assistant professor in 2014 and an associate professor in 2018. Dr. Tombuloglu has more than 18 years of teaching and research experience in genetics, molecular biology, plant genomics, biotechnology, and bioinformatics. His current research is focused on molecular diagnosis, genome and transcriptome sequencing, and nanoparticle–plant interaction. He has published more than eighty internationally indexed articles. He has more than five US patents in molecular diagnosis and biotechnological applications. He is also the editor of five international books.

Guzin Tombuloglu, Ph.D., received an MS in Biology in 2008 and a Ph.D. in Biotechnology in 2014. She has experience in transcriptome sequencing, plant abiotic stress tolerance, and molecular biology of plants. During her Ph.D., she studied transcriptomics identification of boron tolerance mechanism. She has worked on several projects on abiotic stress, plant stress responses, boron toxicity, and transcriptomics. She has taught several courses on genetics, molecular biology, and biotechnology for more than 15 years. She was also chairman of the Pathology Laboratory Techniques Program and assistant manager at the Vocational School of Medical Sciences at the university level. Dr. Tombuloglu has published more than fifty research papers, three books, and several book chapters.

Contents

Preface

With the advent of genomic technologies, our understanding of genetic traits and their possible treatment using genetic engineering methods has improved. These tools can help treat genetic diseases and mutations, as well as develop healthier and more productive organisms. This book is specifically designed to cater to students and researchers in biology, genetics, biotechnology, genomics, medicine, and agriculture across universities, as well as those in postgraduate and graduate programs, and the pharmaceutical and industrial sectors. It highlights the latest developments and discoveries in genomics and biotechnology, as along with open problems and future challenges in producing genetically modified organisms (GMOs).

Readers will be introduced to state-of-the-art developments and trends in gene and genome editing tools and their medical, biotechnological, and industrial applications in different organisms. In addition, the book covers the safety and future applications of GMOs, including country-specific regulations. It also provides examples of how GMOs have contributed to advances in medicine, agriculture, and industry. This book compiles recent findings to meet the needs of its audience.

We anticipate this book will provide readers with an overview of the most recent advancements and trends in GMOs and related research. We thank the contributors for generously accepting our request to share their research and knowledge and for diligently combining their expertise from other sectors to write the chapters. We also express our gratitude to the staff at IntechOpen for their kind assistance throughout the book's creation.

Huseyin Tombuloglu, Ph.D.
Department of Genetics Research,
Institute for Research and Medical Consultations (IRMC),
Imam Abdulrahman bin Faisal University,
Dammam, Saudi Arabia

Guzin Tombuloglu, Ph.D.
Department of Biophysics,
Institute for Research and Medical Consultations (IRMC),
Imam Abdulrahman bin Faisal University,
Dammam, Saudi Arabia

Chapter 1

Perspective Chapter: GMO Foods and Our Health

Tammy Walker-Smith

Abstract

GMO foods come from genetically modified plants that were genetically edited to create healthier and more environmentally tolerant plants that increase nutritional value, resist pesticides, and repel insects. The increase in herbicide uses and the alteration of plant DNA expression has opened the door for harmful toxins expressed in the plant's genetic make-up. These transgenes produce protein-based toxins, such as in corn, which mimic other plant-based toxins that are harmful to humans when consumed. The domino effect of herbicide resistance is that the use of glyphosate (Roundup) increased for commercial purposes because of the resistance to the herbicide. Toxic levels of glyphosate have been found to lead to health problems, including having been identified as a carcinogen-causing agent. The effects of a two-fold alteration of foods genetically and chemically could very well be the Trojan horse of the twenty-first century. Elevated levels of toxins in the food chain may be the link to the declining health status worldwide with higher rates of cancer, diabetes, obesity, Alzheimer's disease, and neurologic conditions. So, how did GMO foods end up being a detriment instead of a benefit?

Keywords: genetically modified organisms (GMO), genetic engineering, crops, glyphosate, CRISPR, toxins, food, health, cancer, food allergy, neurological damage, Alzheimer's, Parkinson's, autism, antibiotic resistance, DNA manipulation, food allergy

1. Introduction

The growing world population is at a cross-roads. One estimation of the earth's population states that it will be around 9 billion people by the year 2050. This increase in population has created a demand for a 70% increase in food production by that time to be able to feed such a large population [1]. Through technological advancements in the GMO field, genetic engineering (GE), and improving crops resistant to infection, viruses, herbicides, and environmental stressors, rates of the most common genetically modified organism (GMO) foods produced have steadily increased by 1.2% annually, but is not at the estimated 2.4% annual increase needed to meet the food demands by 2050 [1, 2].

To better understand what genetically modified organisms are, we need to review the types of GE techniques briefly. Transgene GE is when genes are inserted into a DNA or RNA strand from another species usually from bacterial or mold microorganisms [1]. These specific DNA/RNA sequences control the behavior of the organism/plant through dominant traits. The second category is the transgene-free GMO process that emerged in the early 2000's. This process uses artificial and natural genetic traits to insert gene sequences into an RNA and/or DNA strand for a desired trait. Genomic engineering in the form of cluster regulatory interspaced short palindromic repeats (CRISPR) technology uses insertion and deletion techniques that are more efficient, and cost-effective, but unfortunately are not regulated as stringently due to its ease in the knock in and knock out process of genetic sequences at specific locations. The principle of CRISPR technology has been founded on the action of RNA fragments known as a guide RNA (gRNA) along with a bacterial component that is a CRISPR-associated endonuclease (Cas9) made up of around 20 sequences that makes up the CAS-9 binding "scaffold." This CRISPR/Cas9 creates the platform for the enzyme (cutter) to replace, modify, or delete traits from the DNA fragment and alter its genetic make-up, function, and development [3].

Over the last 20 years, technological advances in Genetically modified organisms have not only improved, but now dominate most processes of how crops, seeds, and plant products impact food production. There will be roughly 11 billion people on the plants by the end of the century, thus projections for the future provide a snapshot of the food production needs/deficit facing the world and the importance of GE foods to achieve food security by then [2]. The premise that the increased crop production will help feed more people in the world and combat food shortages is part of the justification for integrating GMO crops along with increased profits and less impact on the environment. The background for needing more food production stems from estimations of undernourished people to be around 690 million as recent as 2019 [4].

This forward thinking has been the evolution of centuries of farmers cross breeding plants that are better suited to their environment, insect resistant, and mold resistant to improve crop production and variety, farming viability, and crop quality [5]. The only difference today is that this takes place in a laboratory through genetic engineering, thus the changes are no longer naturally occurring, but rather science has incorporated DNA sequence editing to improve plant crops [5]. This chapter will review the impact that such changes to plants and our food has had on the environment and subsequently how these changes are editing the narrative of our health, increases in specific disease processes, increases in autism, exposure to harmful herbicides, increased inflammatory processes, changes in the composition of plants that increase food allergy exposures, and pollination cross-contamination of non-GMO crops [4, 6].

The general population is skeptical of GMO foods since they only have a basic knowledge of what it is and whether it is safe to consume [2, 7]. This dialog controls the narrative and hence has sparked labeling strategies to provide clear symbols or statements of whether the product is a GMO food. This visual cue helps people make informed decisions about what foods they choose to eat. The public acceptance of GMO foods is higher in North America compared to those in the EU [2].

To ensure a product is not GMO based, it should read 100% organic, GMO-Free, or non-GMO project verified. This type of labeling communicates clearly for the consumer to increase transparency and purchasing confidence [7]. One other attempt to control the biosafety of GMO plants resulted in an international agreement called the Cartagena Protocol which was finalized in 2003. This agreement was created to protect the biodiversity of natural plants from the risk of exposure to GMO crops; thereby requiring countries to agree to the entry of specific GMO products/organisms into their country [2].

2. Development of GMO foods

In 1973, the discovery of extraction of DNA from one plant species and the introduction of that DNA into another plant species was the advent of GMO genetic engineering for the first time via laboratory experimentation efforts [8]. This first step in genetic engineering led to the first ever antibiotic resistant crops of tobacco and petunia plants in 1983. This evolution of the science led to a variety of virus resistant tobacco crops in 1990 [8]. The United States approved the first GMO food for human consumption in 1994 which was a tomato variety called the Flavr Savr to increase its shelf life and slow the ripening process. The development of GMO crops such as corn, soy, rice, cotton, and wheat include genetic engineering of specific DNA traits to include greater yields of crops, healthier crops that resist certain molds and bacteria, and herbicide resistant crops to allow for increased herbicide use in the prevention of weed growth in the fields [5, 6]. The highest producer of GMO crops is the United States of America (USA) (see **Figure 1**) with Brazil being the second largest GMO producer. Since GMO engineering began making technological advancements in genetic engineering, the production of these crops have increased 112 fold since 1996 [8]. The time that it takes to bring a GMO product to fruition is around 13 years. This timeline includes studying the impact and/or risks of altering a genetic traits in plants. The average cost for these studies and the development of such products is estimated to be around $140 million USD before it makes it to the market [7]. The rigor that is involved is not conveyed to the general public; however, the unforeseen risks are also not communicated, such as the increased herbicide use on these GMO plants and how that may or may not impact one's health [6, 9].

With great strides in improving crops for productivity and heartier plant strains, genetic engineering took a left turn to make the crops resistant to the herbicide Roundup from Monsanto. This resistance to herbicides created a domino effect that has not stopped since [8]. The ability to use more herbicide on plants to eliminate weeds improved crop production, but also contributed to the development of commercial roundup (Glyphosate) products with additional ingredients that when used together, increase the penetration of the herbicide into the plant that results

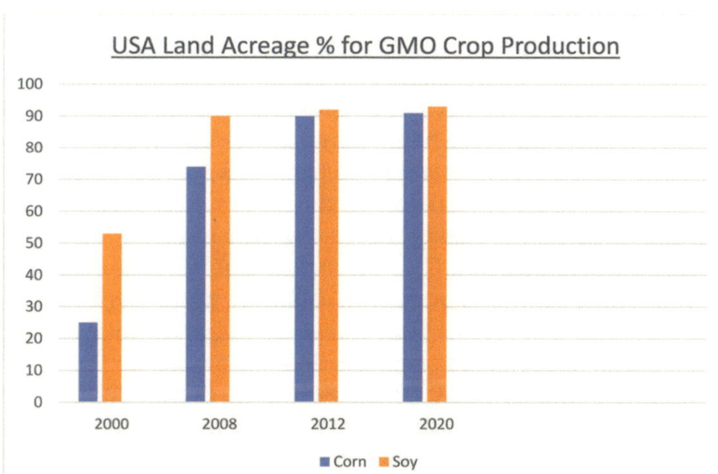

Figure 1.
Percentage of farmland for GMO corn and soy in the USA [1].

in detrimental effects on cellular health in humans. Glyphosate has been classified as "probably carcinogenic to humans" by the International Agency for Research on Cancer (IARC) [6].

3. The detriment of glyphosate in and on GMO foods

To better understand the impact of glyphosate based herbicides (GBH), the amount of this product used annually is around 750,000 tons and is anticipated to increase to as much as 920,000 tons by 2025 [6]. People may ask, if the plant resists the herbicide, then how can it be detrimental? The issue lies in the unforeseen impact on the plant. The advent of this product improved field preparation since there is less tillage needed; however, when you look at the increased use of this product on the crop field, the half-life of the GBH in the soil ranges from 23 to 958 days. This reflects the turnover of crops as much as a three-year period with spraying occurring with each application of crops planted per season. Now, think of the toxic affect and accumulation of this herbicide in the soil that has not reached its half-life and thus provides the foundation for plants to grow and absorb more and more chemically tainted water into the plant itself, not to mention the contamination of aquatic ecosystems from runoff during rain events [10]. This chemical accumulation translates into absorption of toxic levels of GBH within the plant. This has created an unforeseen concern for foods that have elevated levels of GBH along with potential protein-based toxins from genetic modifications of the plant. This combination of exposures has created the perfect storm for chemically induced cellular changes and health issues to humans, animals, and fish due to long-term exposure to these toxins, especially if consumed in high levels (see **Table 1**) [6, 10].

The long-term exposure of glyphosate and its metabolites in foods have been found in vegetables, fruit, cereal/cereal products, rice, beans, and honey [10]. These levels are traced by many European countries for safe levels of the herbicide in foods consumed by the public [10]. The question to ask is: when you eat multiple products daily with "acceptable levels" of glyphosate, what does that translate to with relation to toxicity for daily exposure with extended half-life periods calculated into the equation? Is this why higher levels of multiple disease processes and cancers are increasing?

Exposure to commercial glyphosate products have been found to cause oxidative stress and increases the inflammatory response through the proinflammatory cytokine interleukin-6 (IL-6) genes and tumor necrosis factor alpha gene upregulation in humans [6, 11]. Inflammation is a hallmark to many chronic diseases, the aging process, and cancer [14]. GBH is the trojan horse of this century and no one is the wiser for the impact this chemical has had on the decline of human health and the increase in chronic diseases, the damage caused to the central nervous system, and the decrease in glucose metabolism. Glyphosate is metabolized via the colon and has been found to cause a higher incidence of celiac disease when glyphosate is ingested in addition to other intestinal problems [10]. Is there a link to the increase in cancers and the levels of glyphosate in our foods? This should be explored further.

A systematic review conducted by Costas-Ferria, C.; Duran, R.; and Faro, L. concluded that GBH and other formulations pose "detrimental effects on the human nervous system" which involves Parkinson's and Alzheimer's. A table of their review shows studies that link exposure to GBH and/or its other formulations to an increase in autism, intellectual disabilities, cancer, alterations of glucose uptake in the brain, altered cell death pathways, and increased blood brain barrier (BBB) permeability.

Glyphosate's cellular impact on:	**Humans**	**Rodents/Non-human mammals**	**Fish**
↓ Ribose & ↑ Hypoxanthine leads to ↑ Xanthine and ↑ Uric Acid = Gout	X		
↑ DNA methylation	X		
↑ Inflammatory diseases	X		
Energy metabolism imbalance	X		
↑ Oxidative stress	X	X	
↑ DNA damage	X	X	
↕ Microbiota tolerance	X	X	
Inflammation	X	X	X
↑ Local organ inflammation		X	
↓ Bacterial infection severity		X	
↑ IgG production		X	
↑ Th2 response		X	
↕ Anti-inflammatory cytokines			X
↑ Pro-inflammatory cytokines			X
↓ Antibody production			X
↕ Chemokines			X
↑ Phagocytosis			X
↑ Complement system			X

[6, 11–13].

Table 1.
How glyphosate impacts humans, non-mammals/rodents, and fish on a cellular level.

Glyphosate and its metabolites were found to have induced damage to the central nervous system (CNS) from lactate dehydrogenase (LDH) increases and stating that GBH and its metabolite (AMPA) "reduced the viability of human cells and increased the leakage of LDH" [6]. Most would not think much about LDH increases; but increased levels of LDH have clinical significance for being used prognostically to determine the progression of various types of cancers, is also used as a staging marker for non-seminomatous testicular cancer, and it's levels are high with intracranial hemorrhage or organ damage [15]. GBH was found to cause a decrease in cellular metabolism and induced a "negative regulation in the expression of TUBB3 and GAP43 genes" which are integral in short term memory and memory retention [6, 15]. Could this be the link to Alzheimer's rates increasing? Further studies are needed to support the link definitively between GBH and how it plays a role in neurological/cognitive decline.

Furthermore, the Costas-Ferria et al. noted in their systematic review that commercial glyphosate products used on rodents for research purposes showed increased

Glyphosate and glyphosate based herbicides' impact/effects on human health					
Digestion/Renal	**Reproduction**	**Carcinogenic**	**Cerebral**	**Cardiovascular**	**Cellular**
Microbiota changes	Teratogenicity	↑ Chronic inflammatory syndromes	↑ S100B Levels long-term = Brain damage	Anemia	↑ Oxidative stress
	↑ Autism/Spectrum disorder risk	↑ Risk of cellular ↑ Mutation, thus cancer risk	Memory Issues/Cognitive decline		
Energy metabolism imbalance	Estrogen pathway changes	Non-Hodgkins lymphoma	Crosses the BBB: Neurologic complications	Heart Arrhythmia	Energy metabolism/Altered glucose uptake
Hepatotoxicity & nephrotoxicity		Acute myeloid leukemia	↓ Locomotion		Cytotoxicity = ↓ Cellular viability and ↑ Cell death
			↑ Anxiety/Depression		Genotoxicity

[6, 11–14].
Abbreviations: S100B = Calcium-binding protein, BBB = Blood–brain barrier.

Table 2.
Multiple study findings of how glyphosate/GBH impacts human health negatively.

anxiety/depressive behaviors as well as impairment in working memory, decreased movement/locomotion and decreased sociability patterns [6]. One such study cited in their review found that maternal rodents had a decreased licking behavior with a later increase documented. There was an impairment of neurogenesis and plasticity in the hippocampus. This decrease in remodeling could be the cause for decreased maternal bonding. Decreased neural development of offspring born early was also a finding. Exposure to GBH has been found to modify multiple microRNAs expressed in brain development and linked as a cause of multiple diseases [6].

In summary, GBH exposure over time is linked to alterations in DNA and RNA function as well as impairment of various cellular/metabolic processes, preventing cellular repair, negatively impacting neurological and cognitive functioning, and causing inflammatory triggers that result in chronic diseases and/or cancers (see **Table 2**) [6, 11, 12]. It is important to distinguish between GMO foods and the use of GBH products used on these foods. This distinction reveals that GBH products are negatively impacting health trends worldwide.

4. Plant genome editing challenges

There are multiple ways to edit plant-based genetic sequencing. The traditional way of breeding and crossbreeding plants incorporated cross-pollination of two varieties of the same type of plant that has desirable traits to combine via this route, also known as Mendel's Law of Inheritance [9]. The down side to this type of process is the time it takes to yield a new crop variety after backcrossing over many plant generations to get the end product of a new variety which usually takes a minimum of 10 or more years to accomplish the favorable trait outcomes desired [5]. Transgene-free GMO technology is a challenge in ensuring the safety due to less regulation than transgene GE products since it uses artificial genes in addition to natural genes to alter the genetics of the plant for specific traits to emerge [1]. This brings up the risk to foreign proteins and possible immune responses to this foreign/artificial genetic strain in foods that have been classified safe to eat by the federal food and drug administration (FDA). This issue is highly controversial due to less regulatory monitoring of biotechnology [1].

The issue of antibiotic resistance is of great concern when discussing GE foods and how that translates when looking at long-term health implications to the public [1, 16]. However, now that this concern has been identified, scientists are more careful to not use potential genes that would potentially cause antibiotic resistance. This does not mean that new findings and edited genes will not cause this issue in the future, but scientists have begun to address these precautions to ensure ethical practices going forward in the field of genetic engineering. One caveat to consider in this particular conversation is the focus of antibiotic use in animals raised for human consumption and the risk that poses when super germs are created because antibiotics are no longer effective [16].

Now, the process of GE is more streamlined and has progressed from Zinc Finger DNA-binding domains, although not considered a very efficient process to the newer CRISPR/Cas9 approach that cleaves DNA through an RNA guided genome editing process. This new process is revolutionary in its efficiency for genome editing through cleavage of Cas9-directed DNA sequences that are target and non-target strands called a double-strand DNA break (DSB). This process results in DNA repair that allows for the insertion or deletion (also known as gene knock in or gene knock out) of specific DNA traits through regulation of gene expression by binding DNA or RNA specific sequences [3, 5, 9]. One such example includes the mildew-resistant locus (MLO) gene deleted (knock out) in barley, wheat, and tomato plants to help resist having a powder coated mildew infection that destroys crops. Such knockouts have improved crop viability and resistance to various fungal, bacterial, and viral infections for plants such as rice, cotton, fruits, and vegetables [9]. The ability to protect crops with a natural resistance to such infections allows for healthier plant growth and production of bio sustainable products. This translates into better quality foods, less waste, and more profit for farmers [9].

The long-term effects of this type of genome editing remains to be determined. Better quality crops, yields, infection resistant, and insect, salinity, and drought tolerant varieties have been created in an efficient manner when gene editing isolates and knocks out the weak trait expressed in the plant. The ethical dilemma of GE in plants has bled over into genome editing concerns for human genome alterations that may have long-term consequences such as questionable immune responses and/or targeted gene therapy outcomes [3]. This technology should be considered for humans only once the risk–benefit ratio has been thoroughly assessed [3].

5. GMO food health benefits

When genetic engineering occurs for the good of the public and prevention of malnutrition in areas lacking good nutrition, it can be beneficial. Currently, GMO products make up approximately 66% of the world's diet [2]. There are varieties of rice that are embedded with Beta-carotene in a product called Golden Rice [17]. There have been tomatoes genetically edited to have 4–5 times higher GABA levels. For those with a gluten allergy, researchers have used the CRISPR/Cas9 approach to create a low-gluten wheat for those with celiac disease that have a gluten allergy and are limited in what they can eat, this opens up the door for more food options and safer food selections [8]. Another genome-edited product that the public is benefiting from is lettuce that has increased levels of beta-carotene, thiamine, and vitamin C. Overall, when the focus is improved health and avoiding herbicides, questionable genetic altering practices, and following a code of scientific ethics, the general population benefits greatly from these advances in technology.

Gene editing through the CRISPR/Cas technology has expanded its scope and focus since the advent of fungal, infection, and viral resistant strains of plants. Over the past few years, since CRISPR/Cas technology advancements have occurred, genetic scientists have begun to look to what is possible in improving the health of humans through improving vitamin, mineral, and protein supplementation/amplification with this very same knock out and knock in technique to enrich our foods for better health, decreased risk of cardiovascular and/or cancer risk [4, 9]. Developing countries are benefitting from this technology in the form of biofortified rice with beta-carotene to help reduce/eliminated childhood blindness from malnutrition [1]. One such example is when a tomato plant was genetically altered to increase the bioavailability of Lycopene through the editing of the carotenoid metabolic pathway [9]. Another example includes the increased protein production in barley when d-hordelein was subject to the knockout process, as was a low-gluten transgene-free wheat was developed for those with a high sensitivity to gluten [9]. One major concern to consider for patients with allergies is to understand that the potential for new allergies/sensitivities may occur when a new gene product comes to market such as when a transfer of a protein from one plant to another in merged. However, with CRISPR Cas-9, scientists have been able to remove 35 of the 45 gene responsible for gliadin synthesis (gluten) linked to gluten/wheat allergy and/or celiac disease with an astounding 85% reduction in immunoreactivity [1].

6. GMO risks

The ability to edit plant gene expression to resist herbicides is a risk that most people would not consider. When one thinks of resisting an herbicide, it sounds like less herbicide will be used on the plants; however, quite the contrary, more herbicide is being used to ensure less weeds in the crops. The half-life of glyphosate (Round Up) is a concern because it stays in the soil and accumulates [6, 10]. This accumulation enters the roots of the plant with water and enters the water supply through run off thereby tainting the eco system with lasting effects of Round Up exposure and/or ingestion. The side-effects are discussed at length earlier in the chapter. This trojan horse uses GMO products/plants to make huge profits at the expense of the public [6, 10]

In 1966 the transfer of the Brazil nut protein to soybeans provided a way for the protein from the Brazil nut to be inserted "knocked in" to the soybean, hence this addition of a foreign protein into the soybean triggered allergic reactions from participants that volunteered to evaluate the new product. One major concern to consider for patients with allergies is to understand that the potential for new allergies/sensitivities may occur when a new gene product comes to market such as when a transfer of a protein from one plant to another in merged [4]. When an unfamiliar protein binds to Immunoglobulin E the risk of the protein converting to an allergen in increased. The other risk is that of toxins in new genetically altered foods when a transgene encodes a specific toxin that is now overexpressed [4].

The good news about glyphosate and the health of GMO foods is that glyphosate was to be removed from public home use sales by its new owner Bayer. Glyphosate's previous owner, Monsanto, sold the Roundup company to Bayer in 2018. Bayer stopped selling Roundup to the public in 2023 due to excessive lawsuits regarding glyphosate (Roundup) and its impact on people's health and having caused cancer in those who used it for home use [18]. The sale of commercial glyphosate herbicide has yet to be discontinued from the market amid multi-million-dollar lawsuits. If Bayer's mounting legal troubles are any indication, the commercial formulation may be under scrutiny as well for removal in the commercial sector in the future. One other point to ponder, Bayer is a pharmaceutical company that sells prescriptions for various disease processes, and it owns Round Up which is known to cause various chronic conditions and cancer. Do you think this is a conflict of interest for Bayer to sell the product that causes different disease processes and/or cause damage to humans' health and then turn around and benefit from selling the "treatment/cure?"

7. Conclusion

In conclusion, foods that have undergone insertion/deletion of specific DNA or RNA strands to improve the plant's quality and nutritional value through an ethical and thoughtful purpose poses little threat. The threats lie within the intent and lack of thorough research to determine the risk of human consumption and what that looks like. This includes the exposure to toxic levels of glyphosate, allergens, and foreign components within the altered plant that may be toxic to humans. However, if food products are only released once determined safe for human consumption and are labeled correctly for its modifications and risk for exposure to allergens, then this can be avoided with the appropriate warnings (i.e.: may contain nut protein) [7]. Through trial and error in the development of this new technology, ethical policies and procedures have been initiated to ensure public safety [1, 7]. The time and cost that researchers place on the safety of new products takes years before any GMO products are released for production/planting and processing for foods to be consumed by the public [5, 9].

The public needs to always be weary of the trojan horse from herbicides such as glyphosate (Round Up) and other chemicals that can alter the health, benefit, and purpose of the GMO plant/food [10]. Also, be aware of foods that may increase food allergies due to GE manipulation and cross-contamination of non-GMO plants with GE crops. All scenarios are possible. The takeaway for those wanting to know if it is safe is that there will always be anomalies, but for most GMO products, they are safe to consume. Look for labels that state that the GMO product/food is free from herbicides as well and read food allergy warnings on the package to ensure there is not a risk of food allergy exposure [7]. Knowledge is your best advocate. The more you know, the better equipped you will be to understand the risks and benefits of this new technology and embrace it or make an informed decision to choose/purchase non-GMO products instead [7].

Conflict of interest

The author declares no conflict of interest.

Author details

Tammy Walker-Smith
Texas A & M University Corpus Christi, Corpus Christi, TX, USA

*Address all correspondence to: tammy.walker-smith@tamucc.edu

References

[1] Teferra TF. Should we still worry about the safety of GMO foods? Why and why not? A review. Food Science & Nutrition. 2021;**9**(9):5324-5331

[2] Sendhil R, Nyika J, Yadav S, Mackolil J, Rama Prashat G, Workie E, et al. Genetically modified foods: Bibliometric analysis on consumer perception and preference. GM Crops & Food. 2022;**13**(1):65-85

[3] Moustafa K. CRISPR: Beyond the excitement. Journal of Bioethical Inquiry. Mar 2024;**21**(1):7-9. DOI: 10.1007/s11673-023-10295-y. Epub 2024 Jan 23. PMID:38261130

[4] Lee TH, Ho HK, Leung TF. Genetically modified foods and allergy. Hong Kong Medical Journal. 2017;**23**(3):291-295

[5] Son S, Park SR. Challenges facing CRISPR/Cas9-based genome editing in plants. Frontiers in Plant Science. 2022;**13**:902413

[6] Costas-Ferreira C, Durán R, Faro LRF. Toxic effects of glyphosate on the nervous system: A systematic review. International Journal of Molecular Sciences. 2022;**23**(9):1-28

[7] Ryan CD et al. Exploring the GMO narrative through labeling: Strategies, products, and politics. GM Crops & Food. 2024;**15**(1):51-66

[8] Abdul Aziz M et al. Genetically engineered crops for sustainably enhanced food production systems. Frontiers in Plant Science. 2022;**13**:1027828

[9] Zhang D et al. CRISPR/Cas: A powerful tool for gene function study and crop improvement. Journal of Advanced Research. 2021;**29**:207-221

[10] Soares D et al. Glyphosate use, toxicity and occurrence in food. Food. 2021;**10**(11):1-22

[11] Zhang Q et al. The study of human serum metabolome on the health effects of glyphosate and early warning of potential damage. Chemosphere. 2022;**298**:134308

[12] Andreotti G et al. Glyphosate use and cancer incidence in the agricultural health study. JNCI: Journal of the National Cancer Institute. 2017;**110**(5):509-516

[13] Peillex C, Pelletier M. The impact and toxicity of glyphosate and glyphosate-based herbicides on health and immunity. Journal of Immunotoxicology. 2020;**17**(1):163-174

[14] Guerville F et al. Does inflammation contribute to cancer incidence and mortality during aging? A conceptual review. Cancers. 2022;**14**(7):1622

[15] Farhana A, Lappin SL. Biochemistry, lactate dehydrogenase. In: StatPearls. StatPearls Publishing Copyright © 2024. Treasure Island (FL): StatPearls Publishing LLC.; 2024

[16] Karalis DT et al. Genetically modified products, perspectives and challenges. Cureus. 2020;**12**(3):e7306

[17] Goodman RE. Twenty-eight years of GM food and feed without harm: Why

not accept them? GM Crops & Food. 2024;**15**(1):40-50

[18] Pierson B. Bayer ordered to pay $2.25 billion in latest Roundup trial. Reuters. 2024. Available from: https://www.reuters.com/legal/bayer-ordered-pay-225-billion-latest-roundup-trial-2024-01-26/

Chapter 2

Genetically Modified Crops: A Pivotal Endeavor in Biotechnology

Radhika Velankar, Gauri Nerkar, Mukta Nagpurkar and Kiran Jagtap

Abstract

Transgenic technology has significantly contributed to the genetic improvement of crop plants by improving important agronomic traits like insect/pest resistance, disease resistance, herbicide tolerance, abiotic stress tolerance, and quality improvement. Conventional breeding programs are time consuming and laborious involving screening thousands of progenies for the development of a new hybrid variety. Genetic engineering is a precise tool to develop a new variety in a short duration. Genetically Modified Crops have been used for expression of recombinant proteins of high therapeutic value, monoclonal antibodies, nutraceuticals, edible vaccines, and improved saccharification efficiency of biofuel crops for bioethanol production. The agricultural productivity is limited by global climate changes and unfavorable abiotic and biotic factors posing challenges for crop scientists to meet the rising demand for global food supply. Developing climate-resilient crops will bring more land under agriculture and more vegetation for carbon sequestration thereby annulling global warming. This chapter provides an insight into the principles, advantages, and limitations of the methods used in genetic transformation and the advancements in genome editing, agronomic traits improved in Genetically Modified Crops, potential applications of transgenic technology in biopharming and bioethanol production, biosafety and regulation of transgenic crops, and the challenges in the development of Genetically Modified Crops.

Keywords: Genetically Modified Crops, agronomic traits, biopharming, bioethanol, genome editing, biosafety

1. Introduction

In earlier times 'Biotechnology' was termed as 'the utilization of biological processes, organisms, or systems to produce products that are anticipated to improve human lives'. It is the skill set required for the utilization of living systems or influencing natural processes to produce products, systems, or environments to help human development [1]. The development of rDNA technology circa 1972 shaped the definition of biotechnology into 'the engineering of organisms for the purpose of human usage.' Biotechnology has a wide spectrum of applications in the establishment of hybrid genes and their transfer into desired organisms [1]. The triumphant establishment of rDNA technology has given rise to the terms like Genetically Modified Organisms (GMOs) or Living Modified Organisms (LMOs) or more recently Genetically Edited Organisms (GEOs) [2]. Under the umbrella of 'synthetic biology' an exhaustive compendium of techniques is accessible to scientists for development for gene editing. Green biotechnology articulates the application of biotechnology to agriculture for the development of Genetically Modified Crops (GMCs) [1].

Transgenic crop varieties have been developed not only for increased yield but also for tolerance or resistance to several biotic stress conditions such as insect pests and pathogens, as well as abiotic stress conditions such as high salt, heat, frost, herbicides, drought, and flood. Enhancement or modification of nutritional values, content of several primary and secondary metabolites, natural polymers, and functional biomolecules in various crops has been developed, which are food staples globally. **Table 1** summarizes the biofortification in some of the fruit crops, and **Table 2** summarizes the biofortification in some major vegetable crops [3].

Trivedi et al. [4] described how climate change leads to natural calamities, and in turn diminish natural resources for agricultural practices, food crisis, breach between demand and supply of food and energy as well as high concerns of scientists and policy makers worldwide. Food crisis is growing at an alarming rate as it is getting difficult to keep pace in agriculture production with the rate of population growth. Time-consuming conventional breeding programs are inadequate to deal with such critical situations, especially with the diminishing effects of the green revolution. The inventions of new technologies like genome editing to develop varieties with agronomically desirable, nutritionally significant, and agriculturally suitable traits to provide food security [4].

In light of new challenges due to escalating issues of climate change, global pandemics, and food insecurities, it is essential to build self-dependent, prospering communities in agriculture. GMCs are nifty to bring more land under agriculture, more vegetation for carbon sequestration, and annulling global warming and avert the exodus of climate refugees tussling with hunger and malnutrition. To achieve the same international trade, technology and knowledge transfer should be undertaken without any delay in legislations and approvals to GMCs across the globe irrespective of the country it is developed in or the status quo of the country as developed or

Sr. No.	Name of fruit	Modification in trait	Trade name
1	Apple	Non-browning	Arctic™ Golden Delicious, Granny Smith and Fuji Apples
2	Papaya	Resistance to papaya ringspot virus	Rainbow, SunUp, Huanong No. 1
3	Eggplant	Resistance to eggplant fruit and shoot borer (*Leucinodes orbonalis*)	Bari Bt Begun 1, 2, 3 and 4
4	Pineapple	Delyed ripening/senescence Fruit color Shell morphology	Pinkglow™
5	Squash	Resistance to cucumber mosaic cucuovirus, Zucchini yellow mosaic potyvirus and watermelon mosaic potyvirus 2	CZW3 and ZW20

Table 1.
Biofortification in fruit crops [3].

Sr. No.	Fruit	Trait	Modification strategy	Greenhouse/ field trails	Outcome
1	Apple	Flowering time	Over expression Genome editing	Greenhouse	Early flowering
		Fruit morphology	Over expression Gene silencing	Greenhouse	Different colour, different shape
		Quality improvement	Gene silencing	Field	Increased firmness
		Plant morphology	Over expression	Greenhouse	Smaller trees, dwarf tree
		Disease resistance	Over expression Genome editing	Greenhouse, field	Increased resistance to bacteria and fungi
		Tolerance to abiotic stress	Over expression	Greenhouse	Increased tolerance to drought and cold stress, increased tolerance to salinity
2	Banana	Plant morphology	Genome editing	Greenhouse	Shorter trees
		Disease resistance	Genome editing, gene silencing, over expression	Greenhouse, field	Increased resistance to bacteria and virus
		Nutritional improvement	Genome editing	Greenhouse	Increased carotenoid content
3	Blueberry	Flowering time	Over expression	Greenhouse	Early flowering
4	Cherry	Fruit morphology	Gene silencing	Greenhouse	Smaller fruits
		Disease resistance	Gene silencing	Greenhouse	Increased resistance to virus
5	Citrus rootstock species	Plant morphology	Over expression, Gene silencing	Greenhouse	Shorter trees
		Disease resistance	Over expression	Greenhouse	Increased resistance to bacteria
		Tolerance to abiotic stress	Over expression	Greenhouse	Increased tolerance to drought stress
		Flowering time	Over expression	Greenhouse	Early flowering
6	Orange	Disease resistance	Genome editing, over expression, down-regulation	Greenhouse	Increased resistance to bacteria
		Nutritional improvement	Gene silencing	Greenhouse	Increased carotenoid content
7	Grapefruit	Disease resistance	Over expression	Greenhouse	Increased resistance to fungi
8	Lime	Disease resistance	Over expression	Greenhouse	Increased resistance to virus
9	Cucumber	Disease resistance	Genome editing	Greenhouse	Increased resistance to virus
10	Grapevine	Disease resistance	Over expression	Field	Increased resistance to virus

Sr. No.	Fruit	Trait	Modification strategy	Greenhouse/ field trails	Outcome
		Fruit morphology	Over expression	Field	Reduce pathogen-induced mortality
		Tolerance to abiotic stress	Over expression	Greenhouse	Different colour, increased tolerance to salinity and cold stress
11	Kiwi	Nutritional improvement	Over expression	Greenhouse	Increased carotenoid content
		Quality improvement	Gene silencing	Greenhouse	Ripening
		Tolerance to abiotic stress	Over expression	Greenhouse	Increased tolerance to salinity
12	Melon	Disease resistance	Gene silencing	Greenhouse	Increased resistance to virus
13	Papaya	Quality improvement	Gene silencing, gene silencing	Field	Delayed fruit ripening
14	Pear	Quality improvement	Over expression	Greenhouse	Decreased ethylene production
		Disease resistance	Over expression	Greenhouse	Increased resistance to bacteria
		Nutritional improvement	Over expression	Greenhouse	Increased tocopherol content
15	Pepper	Disease resistance	Over expression	Greenhouse	Increased resistance to fungi
		Tolerance to abiotic stress	Over expression	Greenhouse	Increased tolerance to salinity
16	Plum	Flowering time	Over expression	Greenhouse	Early flowering
		Disease resistance	Gene silencing	Greenhouse	Increased resistance to virus
17	Strawberry	Flowering time	Genome editing	Greenhouse	Early flowering
		Nutritional improvement	Gene silencing	Greenhouse	Decreased starch and increased soluble sugar content, increased anthocyanin content
		Quality improvement	Over expression, genome editing	Greenhouse	Increased fruit firmness
18	Tomato	Flowering time	Genome editing	greenhouse, field	Early flowering
		Quality improvement	Genome editing	Greenhouse, field	Increased shelf-life
		Fruit morphology	Over expression, genome editing	Greenhouse	Parthenocarpic fruits
		Nutritional improvement	Genome editing	Greenhouse	Increased lycopene content
		Disease resistance	Over expression, genome editing	Greenhouse	Increased resistance to bacteria
19	Walnut	Insect resistance	Over expression	Greenhouse	Increased resistance to insect

Sr. No.	Fruit	Trait	Modification strategy	Greenhouse/ field trails	Outcome
20	Watermelon	Pest resistance	Genome editing	Greenhouse	Increased herbicide resistance
		Disease resistance	Gene silencing	Greenhouse	Increased resistance to virus

Table 2.
Biofortification in vegetable crops [3].

developing [5]. For the development of a global sustainable community, there is a dire need of a growing body of quantified peer-review literature on economic, environmental, and health benefits and educated perspectives of farmers and other stakeholders on adoption of GMCs across globe [5]. Additionally, fierce campaigns of misinformation about the dangers and hazards of GM crops by environmental non-governmental organizations without any scientific assessment of claims and accusations merely for their political and policy influences should be drawn to a close end [5]. The regulatory delays for the approval of new GM crops and frequent international commodity trade failures or challenges lean a detrimental impact on improving food security and other enormous benefits of GMCs [5]. The driving force behind international acceptance pertaining to GM crop approval, adoption, and trade is less scientific and more political. If more scientific factors determined the benefits of GM, then opposition to GM could have been diminished in past decades [5]. This chapter provides insight into the principles, advantage, and limitations of the methods used in genetic transformation; agronomic traits improved in Genetically Modified Crops; the potential applications of transgenic technology in biopharming and bioethanol production; and a brief note on biosafety and regulation of transgenic crops. The challenges and perspectives pertaining to the Genetically Modified Crop development are also discussed.

2. GMCs: role of perception

Lucht [6] scrutinized the vast gap between the acceptance of GMCs by farmers and acceptance by consumers globally. It is essential to focus on common aims, underlying values, and reasons for the adoption and acceptance of GMCs by famers. The meta-analysis of 147 agronomical studies showed that GMCs perform differently in different world regions and in different agricultural systems. Farmers benefited from increases in yield and profit, while additional non-monetary benefits, such as time savings, ease of use, and more flexibility in their planning were accompanied. The higher seed cost for biotech varieties can be redeemed by pronounced reduction in insecticide quantities due to insect-resistant Bt-crops and the possibility to switch to more environmentally benign herbicides with herbicide-tolerant crops. The vital role of GMCs as feed in many countries like USA is also mentioned [6].

There are obvious advantages of GM crops for farmers as they address their specific needs. The perceptions and discernment of consumers are pivotal in acceptance, wide spread cultivation, and adoption of GM varieties in agriculture. Insect resistance or herbicide tolerance is beneficial traits for the farmers but have no perceptible influence on food characteristics or quality, which will be evaluated by consumers [6]. Mustapa et al. [7] mentioned GMCs as a controversial global issue since their commercialization in 1996 and carried out a study to analyze farmers' attitudes toward GM crops in Malaysia. A descriptive study concluded that the farmers have a high level of self-efficacy and perceived high benefits of GM crops but are also disappointed due to the low level of support from the government [7]. Zhaleh et al. [8] analyzed factors affecting consumer's favorable reception of GMCs as industries or policymakers using a case study on edible oil. The analysis insists on securing the trust of people by shedding light on the production process and scientific debates to discuss various aspects of GMCs [8]. Rahnama et al. [9] performed compositional and morphological analysis of salt stress-tolerant mannitol-1-phosphate dehydrogenase (mtld)-transgenic potato plants. The study was significant to infer that 'this variation is a common phenomenon among potato varieties and substantial equivalence between transgenic and non-transgenic varieties [9]. Bekele-Alemu and Ligaba-Osena [10] explored *in silico* approach to reveal drought-responsive genes in distinct tissues of plants like rice, wheat, maize, barley, sorghum, pearl millet, and the model plant Arabidopsis, which were found to be promising candidates for trait improvement in crops using transgenics [10]. Hansson [11] emphasized a scientific approach in the design of precautionary measures to protect human health and the environment. Advanced knowledge in genetics, plant biology, and ecology can endow us with much better techniques and parameters to analyze the risk imposed due to widespread use of GMCs. Relevant scientific advances should be directed at potential danger with credible scientific evidences [11].

2.1 Example of GMCs: the Arctic apple

The Arctic® apple is developed using a sense-posttranscriptional silencing approach, using a chimeric polyphenol oxidase (PPO) gene targeted to disrupt the expression of the polyphenol oxidase enzyme and prevent unsightly browning in the fruit flesh. Browning of flesh is one of the prominent causes of food wastage in the postharvest process.

Waltz [12] mentioned a growing trend of pre-cut fruit and vegetables in the food industry. The Arctic apple will enable food companies to cut and package the apples without adding browning inhibitors like calcium ascorbate, making the trait useful to food industries as well [12].

3. Tools and techniques: plant transformation methods

Genetic transformation of crop plants mainly employs direct and indirect methods. Agrobacterium-mediated gene transfer is an indirect genetic transformation method as it uses organisms as a vector to introduce the foreign DNA into the target cells. Direct genetic transformation methods use external forces to deliver target genes into plant cells, including particle bombardment/gene gun, electroporation, liposomes, silicon carbide, microinjection, and pollen-tube-pathway-mediated plant genetic transformation methods. In this chapter, we are going to discuss two methods, namely *Agrobacterium*-mediated transformation and Biolistic transformation/particle bombardment, which are the most applied methods in genetic transformation of crop plants.

3.1 *Agrobacterium*-mediated transformation

Agrobacterium tumefaciens contains (Ti) tumor-inducing plasmids which contain a T-DNA region that integrates with the host genome. The Ti plasmid can be modified to incorporate a target gene in T-DNA region, such that upon transformation of the plant cells with this modified Ti plasmid, the target gene can be co-integrated with the T-DNA into the plant genome. The injured part of the plant releases phenolics or acidic sugars which are sensed by VirA, leading to the activation of VirG via phosphorylation. Following the induction of VirD1 and VirD2, the T-DNA region is cleaved at the left border (LB) and right border (RB) repeat sequences in the Ti plasmid. VirD2 attached to the 5′ end of the T-DNA (VirD2/T-DNA) leaves the bacteria via the Type IV secretion system. Furthermore, VirD2/VirE2/T-DNA T-complex promotes T-DNA integration into the plant genome [13].

Agrobacterium-mediated plant genetic transformation has become the most widely applied tool for the genetic transformation of crop plants because of its high efficiency, simple method, genetic stability of the transgene, and amenability of the majority of dicots and a few monocot plants for transformation. However, the major bottleneck in the application of *Agrobacterium*-mediated transformation in monocot plants is that most monocots cannot be naturally infected by *Agrobacterium* [14]. Nevertheless, understanding the key requisites for effective infection and gene transfer, such as using immature embryos that have a large number of actively dividing cells; efficient selection stably transformed cells from a large number of non-transformed cells; uing herbicide-resistant markers; and use of hyper-virulent *Agrobacterium* strain, has paved the way for the development of transformation protocols, as well as the use of suitable selectable markers [14].

3.2 Particle bombardment/biolistic transformation

The advent of biolistic transformation technique has paved the way for the transformation of important agronomic crops such as maize [15], soybean [16], and cotton [17], which were not attainable by *Agrobacterium*-mediated transformation. Particle bombardment by gene gun is a physical method for incorporating foreign DNA into the plant genome, which is not limited to specific target tissue but is amenable to a host of different types of cells such as calli, immature embryos, and organs. A DNA-coated microcarrier is created by coating the target DNA on the surface of gold or tungsten particle. Using an electric discharge or a pressurized helium gas stream, the DNA-coated microcarrier is accelerated by high-pressure helium into the gas acceleration tube. The high-pressure created in the vacuum chamber accelerates the particle to gain sufficient momentum to pierce recipient cells at high speed. Thus, the target gene coated on the surface of the particles enters the cell [18, 19] and is eventually integrated into the plant's chromosome, producing the transgenic plant [20].

Biolistic transformation can be used for diverse cell types. Although embryogenic callus is often preferable, organized tissues such as embryos or meristems can also be successfully produced using somaclonal variations observed in tissue culture [21]. However, biolistic transformation has a lower transformation rate, is highly expensive, and uses unprotected exogenous DNA [22]. DNA fragments smaller than 10 kb are preferred for biolistic transformation, as larger fragments may break during the bombardment process, creating a problem for expression of the transgene as it loses its integrity [18].

3.3 Genome editing tools

Recent advances in the field of genetic engineering have enabled genetic engineers to precisely modify plant genomes. Lack transgene silencing, the absence of foreign

DNA in the genome-edited plants, and the faster and cheaper method of genome modification are a few remarkable features that have resulted in its widespread adoption of genome editing in crop breeding in less than a decade [23].

Sequence-specific nucleases facilitate the precise and targeted modifications of the complex genome soma clonal editing. SSNs are engineered endonucleases that generate DNA double-strand breaks (DSBs) in the genome, resulting in site-specific insertion, deletion, and substitution of the DNA. SSNs are programmed to recognize preselected genomic sites and they make use of cellular DSB repair mechanisms such as non-homologous end joining (NHEJ) or homology-directed repair (HDR). In NHEJ, random insertion or deletion of DNA at the cut site before reattaching of the free DNA ends renders a gene non-functional [24]. In HDR, a Donor DNA of choice is added which is homologous to the site of the break. Cells use this as a patch to repair the DNA [25]. HDR facilitates the introduction of a new gene with a vital function or correcting a mutation by replacing the mutated sequence with a healthy sequence [26]. Based on the ability to recognize the target DNA sequence within the complex genome, SSNs have been classified into four major classes: *viz.* engineered homing endonucleases or meganucleases (MNs), zinc finger nucleases (ZFNs), transcription activator-like effector nucleases (TALENs), and clustered regularly interspaced short palindromic repeats (CRISPR) and CRISPR-associated (Cas) nuclease 9 (CRISPR/Cas9).

Meganucleases have target sites of up to 18 bp [27]. ZFNs were first reported by Kim et al. as sequence-specific chimeric proteins containing a DNA binding domain fused to a non-specific cleavage domain (derived from type II restriction enzyme *Fok*I) [28]. The DNA binding domain consists of 3–6 Cys2-His2 tandemly arranged zinc finger repeats that recognize 9–18 bp sequences (3 bp by each ZFN unit) [29]. Altering the DNA binding domain and catalytic domain through mutagenesis can facilitate the development of custom-designed zinc fingers (ZnF) [30].

TALENs consist of an endonuclease domain which pairs with multiple transcription activator-like effector domains that recognize single base pairs [31]. TALENs are virulence proteins derived from the plant pathogen *Xanthomonas* sp. consisting of a DNA binding domain and *Fok*I nuclease domain. These domains act as dimers, binding to opposite strands of DNA, separated by a spacer sequence of 12–20 bp, thus creating a double-stranded break. The DNA binding domain consists of 33–35 tandemly arranged amino acid repeats [32]. These repeats are similar except for two highly variable amino acids at positions 12 and 13, called repeat variable di-residue (RVD), which are responsible for specific base recognition and engineering of these bases in repeats [33]. A total of four RVD modules can recognize each of the bases: guanine (G), adenine (A), cytosine (C), and thymine (T), and each module is able to function independently.

The most recent ground-breaking technology for genome editing is the type II clustered regularly interspaced short palindromic repeat (CRISPR)/Cas9 (CRISPR-associated) system from *Streptococcus pyogenes* [34]. CRISPR/Cas9 achieves targeting through a guide RNA (sgRNA), which are short nucleotide sequences (~20 nt) with a specific sequence that can target the genomic sequence of interest. The Cas9 nuclease then cleaves the resulting RNA/DNA complex, creating a DSB at the target site containing a conserved protospacer adjacent motif (PAM). Repair occurs by NHEJ, which creates indels in the protein-coding regions causing frameshift or knockdown of the desired genes [35]. The simplicity of DNA targeting through base-pairing has led to the quick and broad adoption of CRIPSR/Cas9 reagents for genome editing.

4. Development of transgenic crops for important agronomic traits

4.1 Production of transgenic plants resistant to abiotic and biotic stress

Plants growing in the environment are exposed to many biotic and abiotic stresses such as saline stress, pesticide exposure, microbial population present in soil, global warming, and the presence of environmental pollutants [36]. As a result of this, plants have developed elaborate mechanisms to perceive, respond to, and adapt to stressful situations. Plants have multiple molecular signaling pathways, such as up-regulation of stress resistance genes, and production of metabolites like proteins which can handle adverse situations. In many cases, plants show a similar response to different environmental stresses [37]. Plants respond to cold, salt, and drought stresses by expressing similar genes share similar genes, which can be used by the plants in these stresses [38]. The interaction of different stressful events can have adverse effects on plant growth [39]. Drought stress, can weaken the plant's defense system, can lead to stagnant plant growth, or can encourage pathogen proliferation in the environment [40]. **Figure 1** depicts the response of plants to different types of biotic and abiotic stress.

Conventional plant breeding practices, although improving crop yield, are not very efficient in enhancing the tolerance of plants to abiotic stress [41]. Hence, it becomes mandatory to adopt alternative approaches which can be used to improve tolerance capabilities in plants to combat biotic and abiotic stresses. The International Service for the Acquisition of Agro-Biotech Applications (ISAAA) reported that India, America, Brazil, Argentina, and Canada were the top five countries that produce genetically engineered crop varieties. In 2019, transgenic crops like soybean, maize, cotton, and canola were mainly produced by these countries [42]. Thus, continuous attempts are being made to enhance the performance of the plants in stress conditions and to increase the plant yields. This has proved advantageous for

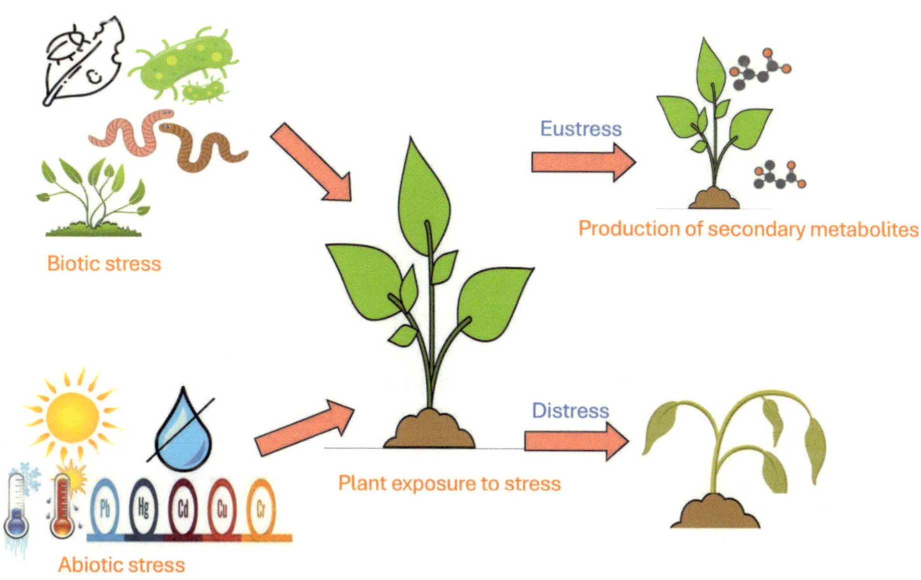

Figure 1.
Reponses of plant to different types of stress.

humanity and for addressing the increasing food demand to feed the increasing population. The currently used advanced techniques like CRISPR/Cas9 also helps in selection of plants which can tolerate stress conditions [43].

Rice which is considered as one of the important food crops, is constantly exposed to various biotic stress such as bacterial, fungal, viral, pests, and nematode infections. Hence, there is an utmost need to develop biotic stress resistance rice variety. Genetic engineering technologies like RNAi and the creation of transgenic rice have contributed to the development of resistance rice varieties for different bacterial, fungal, and viral diseases. Transgenic technology inserts resistance genes which are isolated from organisms like plants, animals, and microbes [44].

Transgenic potato lines were developed to increase the salt tolerance by introducing a pyrroline-5-carboxylate synthetase (P5CS) gene from *Arabidopsis thaliana* using *Agrobacterium*-mediated gene transfer method. This enzyme converts glutamate to D1-pyrroline-5-carboxylate, which is then reduced to Proline. Proline is an osmoprotectant that protects the cell structures under osmotic stress. The produced transgenic potato showed enhanced proline production as compared to normal potato crops. Proline accumulation notably increased in the presence of a 100 mM salt concentration. This transgenic plant did not have much altered tuber yield and weight [45].

Many reports on the transformation of sugarcane crops to make it resistant to biotic and abiotic stress have been published [46, 47]. The success of creating transgenic sugarcane highly depends upon some important factors like method of transformation, types of promoters used in the vectors, the nature of the target tissue or explant, the selection system used, and the system to regenerate the tissue. Sugarcane is a glycophyte and the drought and salinity stress have a severe impact on its development and sucrose contents. Using bioengineering approaches, the Arabidopsis Vacuolar Pyrophosphatase (AVP1) gene, which confers drought and salinity stress resistance, was introduced into sugarcane using *Agrobacterium*. Apical buds of the sugarcane variety CP77-400 were used as explants. *Agrobacterium* was used for transformation with the pGreen 0029 vector had the introduced AVP1 gene under the control of 35SCaMV promoter. Addition of acetosyringone, and some antibiotics like cefotaxime, kanamycin is required for successful transformation. A regeneration frequency of 77.5% was obtained on MS media. The results of the experiment highlighted the fact that apical buds can be considered as feasible tissues for transformation of sugarcane, which produced many transgenic plants in less time. The expression of the transgene was detected using RT-PCR. Only the transgenic plants were able to grow in higher salt stress and water scarcity, hence exhibiting resistance against salinity and drought stress [48].

Genetic engineering of plants to develop resistance to biotic stress is an efficient method pathogens and pests. This technology is specific and targeted, which ensures that only the necessary effects are visible [49]. Genetic transformation technology thus helps to transform an unrelated gene in plants or organisms, while genome editing helps to change, insert, or replace a specific gene sequence within the genome [50]. Some genetic engineering experiments done in different plants to make them resistant to abiotic and biotic stress are listed in **Table 3**.

4.2 Use of plant biotechnology in increasing oil content of crops

Oilseed crops are considered as an integral part of human diet and are slowly replacing conventional fossil fuels to meet energy needs. From past many years, many remarkable events in genetic engineering to enhance oil content in the oilseed crops like rape, camelina, soybean, and maize [51]. Many studies have reported in the past

Stress type	Trait	Transferred gene	Gene function	Plant	Transformation method
Biotic	Herbicide resistance	EPSPS	Glyphosate resistance	Sugarcane	Gene gun
	Pest resistance	synthetic δ-endotoxin *cryIA(b) gene*	Resistance to the European Corn Borer (ECB)	Maize	Biolistic
	Disease resistance	β 1,3-Glucanase	Degradation of fungal cell wall	Sugarcane	*Agrobacterium*
Abiotic stress	Drought resistance	AVP-1	Osmotic regulation	Sugarcane	*Agrobacterium*/gene gun
	Salinity	P5CS	Proline synthesis	Sugarcane	Biolistic
	Salinity	OVP1	Enhanced salt tolerance, chlorophyll content and membrane stability	Rice	Gene gun
	Salinity	P5CS	Improved salt tolerance and better root growth	*Vigna aconitifolia*	Particle bombardment
	Salinity	TsVP	Enhanced salinity stress, reduced membrane leakage, and increased photosynthetic performance	Cotton	Particle bombardment
	Salinity	P5Cs	Increased proline content and salt tolerance	Potato	*Agrobacterium*-mediated
	Salinity	SeVP	Enhanced salt tolerance	Wheat	Plasmid based

Table 3.
Genetic Modifications to create biotic and abiotic stress resistant crops.

discussing transgenic technologies used for enhancement of oil content of seeds, which includes manipulation of Triacyl glycerol or fatty acid synthesis, and modification of carbon lux toward triacylglycerol synthesis, and introduction of new pathways for synthesis of triacylglycerol taken from mammalian or yeast system [52]. Many new advanced methods like monomeric transcription activator-like effectors-based nucleases (mTALENs) and modified forms of the clustered regularly interspaced palindromic repeat (CRISPR/Cas 9 system) are applied for improvement of oil crops in terms of yield and nutritional quality of oils (**Figure 2**) [53, 54].

In animals, the monoacylglycerol acyltransferases (MGATs) are associated with lipid absorption and resynthesis in the intestine. The first step of catalysis in the monoacylglycerol (MAG) pathway by acylating MAG to form diacylglycerol (DAG) takes place with the help of this enzyme. In plants, the Kennedy pathway of triacylglycerol (TAG) biosynthesis does not involve MGAT catalysis step. The DAG and TAG are synthesized from glycerol-3-phosphate (G-3-P) by three subsequent acylation reactions. The heterologous expression of mouse MGAT acyltransferase in *Nicotiana benthamiana* increased accumulation of TAG in vegetative tissues. The DAG produced by MGAT acyltransferase

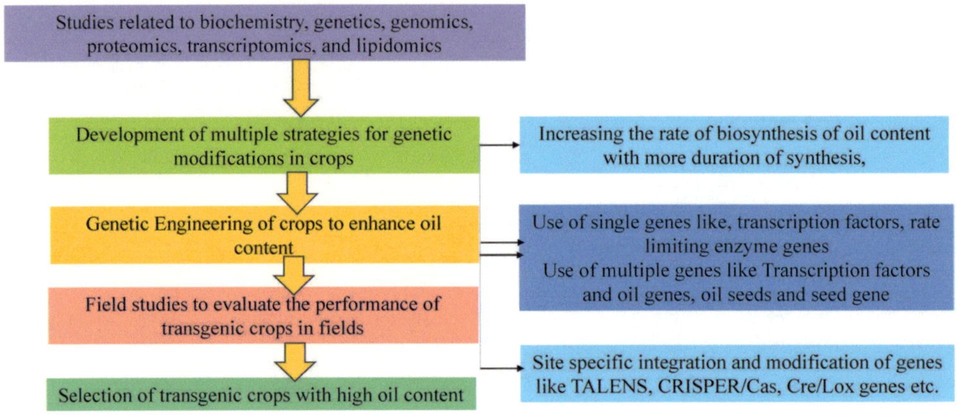

Figure 2.
Strategies for genetic engineering in crops for enhancement of oil content in seeds.

can serve as a substrate for native and co-expressed diacylglycerol acyltransferases (DGAT). This study demonstrates a new concept of increasing oil content in vegetative tissues by using MAG as a substrate for TAG biosynthesis was demonstrated [55].

Triacylglycerol (TAG) can be produced from high biomass crops using genetic engineering approaches. TAG usually accumulates in plant seeds after germination, while its production in vegetative tissues is less. The accumulation of 15% TAG in *Nicotiana tabacum* (tobacco) leaves was reported by the simultaneous expression of three genes involved in different steps of TAG production without any negative impact on the plant development. Mass Spectrometry and Confocal fluorescence microscopy were used to confirm the accumulation of TAG within the leaf mesophyll cells [56]. The genetic modifications done in crops to increase their oil content is depicted in **Table 4**.

4.3 Biofortification

Biofortification is the process of increasing and concentrating the available micronutrient in crop plants through breeding or genetic engineering. Biofortified crops can be used to improve human nutrition and have the potential to provide the micronutrients required in remote communities with low diet variability.

Biofortified crops have bright prospects in addressing the malnutrition challenge of the world [57]. Rice is a food staple for more than 3.5 billion people around the world

Transferred genes	Plant	Oil phenotype
SDP1	*Arabidopsis*	Increase in oil seed by 3.1%
WR11, DGAT1	*Arabidopsis*	Increase in oil seed by 6%
MGAT 1	*Nicotiana benthamiana*	Increases in leaf TAG level 7.3 times
PDAT, OLE1 and WRI1	*Arabidopsis* (tdg5)	Increase oil content by 8.5% of leaf dry weight
BnGPDH, BnDGAT, BnGPAT and ScLPAAT	*Brassica napus*	Increase in seed oil content by 14.40%

Table 4.
Genetically Modified Crops for increased oil content.

and therefore an ideal target crop to address the global challenge of undernutrition. The underprivileged population is severely affected by vitamin deficiency due to poor affordability. An important breakthrough in this direction was the development of Golden Rice enriched with provitamin A (beta-carotene) [58]. By targeting gene encoding carotene desaturase, phytoene (β-carotene precursor) level has been enhanced by up to 23-fold in the transgenic rice lines [58]. Also, by overexpressing genes encoding Arabidopsis GTP-cyclohydrolase I (GTPCHI) and aminodeoxychorismate synthase [ADCS] [59], the folate content in rice has been increased up to 150-fold. Thus, the daily folate requirements of an adult can be met by consuming 100 g of modified rice.

Wheat is another important staple food crop in the world which makes it an ideal candidate for addressing the challenges of most nutrient deficiencies like vitamin A, iron, and quality proteins. Enhancement of the provitamin A content by expressing bacterial PSY and carotene desaturase genes [60] iron content expression of ferritin gene from soybean [61]; protein content by expressing *Amaranthus* albumin gene [*ama1*] [62], antioxidant activity by expressing maize regulatory genes (C1, B-peru) involved in anthocyanin production [63]. Wheat has also been targeted to address the challenge of overnutrition and obesity, by silencing gene encoding SBE [SBEIIa] [64] to enhance the content of less digestible and resistant amylose starch.

Maize, an important staple crop in developing countries, has been targeted for increase vitamins, minerals, quality protein, and reduction of antinutrient components by means of genetic engineering. The zinc content in barley has been improved by overexpression of zinc transporters [65]. Sorghum is an important staple of poor rural people. Sorghum's seed storage protein, γ-kafirin, is resistant to protease digestion making it less digestible than the other major staple crops. The transgenic sorghum lines created by RNAi silencing of the γ-kafirin [66] have improved digestibility index.

Being a global source of vegetable oil and high-quality protein, soybean has been genetically engineered to increase provitamin A (beta-carotene), a monounsaturated ω-9 fatty acid (oleic acid) and seed protein contents by expressing bacterial PSY gene [67]. Potato is the world's fourth most important source of calories, and any nutritional enhancement is of great significance. Provitamin A (carotenoid forms) has been increased by simultaneous incorporation of three genes: PSY, phytoene desaturase, and lycopene β-cyclase [68] in potato. As carrots are poor in calcium content, attempts have been made to increase bioavailable calcium content in transgenic carrot by expressing the Arabidopsis H^+/Ca^{2+} transporter [CAX1] [69]. In tomato, higher contents of lycopene, beta-carotene, and lutein have been achieved by the expression of PSY gene [crtB] [70].

The antioxidant capacity of apple has been expanded by introducing the stilbene synthase gene from the grapevine (*Vitis vinifera* L.) to increase the biosynthesis of resveratrol in transgenic apple. Banana has been predominantly targeted for beta-carotene [71]. The content of the sulfur-containing amino acid, methionine, has been increased by the expression of cystathionine γ-synthase [AtCgS] [72] in Alfalfa.

4.4 Improvement of photosynthetic efficiency in transgenic crops

Wang et al. [73] threw light on evident fact of rising atmospheric carbon dioxide concentration stimulating rice yields with a ubiquitous decline in nutrition like protein, iron, and zinc. The two GM rice lines, one with an enlarged root system another with enhanced nitrate absorption, demonstrated both greater yield enhancement and better nutritional quality at elevated carbon dioxide relative to their non-GM counterparts. The enlarged root system GM variety could enhance general nutrition protein and micronutrients and enhanced nitrate absorption. GM varieties exclusively improved protein concentration under elevated carbon dioxide [73].

Kumar et al. [74] discussed the commercial cultivation of microalgae adversely affected due to inefficient photosynthetic efficiency from limited biomass yield. An effective mechanism to enhance photosynthetic efficiency and overall biomass productivity in microalgal cultures is enhanced by minimizing the light-harvesting antenna size of the photosystems. The strategies like mutagenesis, through UV radiations and chemical mutagenesis, genetic engineering, and DNA insertional mutagenesis have been applied to obtain mutant strains possessing a regulated antenna with a regulated limited number of light-harvesting molecules. The recent developments in truncated antenna mutants of microalgae, aiming to increase the photosynthetic efficiency and biomass productivity of the respective cultures were discussed. For example, *Chlorella pyrenoidosa*, *Chlorella sorokiniana*, *Chlorella vulgaris*, and *Chlamydomonas reinhardtii* [74].

In order to meet increased food and energy demands of humankind, photosynthetic activity has to be increased. To increase photosynthetic efficiency targets are decreasing photosynthetic antennae size, increasing the photosynthetically available light spectrum, countering oxygenase activity of Rubisco by implementing C4 photosynthesis to C3 plants and altering source to sink transport of metabolites. The proposed targets such as sugar alcohol metabolism and root to shoot carbon dioxide transport are under consideration to improve photosynthetic performance and drought tolerance at the same time [75].

4.5 Carbon credits

Brookes reviewed environmental impacts due to adoption of GMCs globally. For example, adoption of GM insect-resistant and herbicide-tolerant crops has reduced pesticide spraying by 775.4 million kg (8.3%), resulting in decreased the environmental impact associated with herbicide and insecticide use on these crops [76]. Sutherland et al. [77] highlighted that earlier tillage was the way to control weeds but now due to GMCs minimum or zero tillage is possible for continuous cropping without soil disturbance. GMCs make transition of farmland as being a net carbon emitter to being a net carbon sequestrator. The study revealed the correlation between genetically edited, herbicide-tolerant crops and glyphosate use is a driver of the increased soil carbon sequestration. The removal of tillage and adoption of minimal soil disturbances has reduced the amount of carbon released from tillage and increased the sequestration of carbon through continuous crop production [77].

Crop production and productivity increased significantly during the era of the adoption of GM crops; some of this increase can be attributed to GM technology and the yield protection traits that it has made possible even if the GM traits implemented to-date are not yield traits per se. GM crops have also been credited with helping to improve farm incomes and reduce pesticide use [78].

5. GM crops for value added products

5.1 Bioethanol production from Genetically Modified Crops

Plants are considered as a cheap, and highly efficient energy producer since it converts light energy into simple sugars using photosynthesis and carbon dioxide fixation. The sugars formed are then used to produce combustible polymers and help in generation of composite cellulosic secondary cell wall, hemicellulose, and lignin [79]. Recent advances in the field of plant biotechnology have proved very useful in the production of bioethanol. This helps increase the quantity of biofuel production from plants without a proportional increase in the cultivable land required for cultivation of the plants. The development of genetically modified plants for increased yield of bioethanol can probably be the most efficient and quick solution to meet the ever-increasing energy demands, especially for the ethanol produced from lignocellulosic biomass [80].

There are many ways to customize the biomass content of plants to obtain a good bioethanol yield including traditional plant breeding practices or conventional molecular biology techniques. However, apart from these techniques, the role of genetic engineering is also important in modification grass biomass content to produce optimum bioethanol from it. The most common methods for transferring a foreign gene into plants are the biological method using *Agrobacterium tumefaciens* and the physical method of gene gun [81]. Many researches in the past as well are in the present indicate use of different genetically modified plants as a potential source for bioethanol production and some of them are discussed below. In the future, the production and usage of ethanol that is obtained from the cellulose of grains can decrease the dependence on the use of petroleum [82]. The corn variety with enhanced starch content can be used as a good source of biofuel [81]. The corn grains from these genetically modified varieties may result in increased glucose levels which eventually leads to high ethanol production. These crops which are modified to get high ethanol yield are usually also modified in such a way that they show resistance against biotic and abiotic stresses like insect attacks, herbicides, drought, and salinity. As, when these traits are expressed in the plants, then only the farmers can make his cultivation practice successful and affordable [81]. Different approaches to enhance bioethanol production from plants includes production of transgenic cellulase, modification of cell wall and lignin content, and increase in starch and sugar content. There are different generations of biofuel production depending upon the sources from which the ethanol is produced.

5.1.1 First generation biofuels

The bioethanol produced mainly from edible plants belongs to the first generation [83]. The sugarcane (*Saccharum officinarum* sp.) crop varieties that are used nowadays for biofuel production are hybrids. To increase the sucrose content of sugarcane, a gene from bacteria was introduced in sugarcane. The gene, sucrose isomerase converts sucrose into isomaltulose, one of its isomers [84]. The produced isomaltulose was allowed to accumulate in the vacuoles, and the accumulated sugar was almost twice the amount of total sugar found in control plants. The producer plant did not metabolize the isomaltulose, which could be used for bioethanol production. Corn derived production of bioethanol has got demand after development of pest and herbicide-resistant corn varieties [85]. The starch biosynthesis pathway has been manipulated in corn varieties so that the prepared starch can gelatinize at less temperature and is more accessible to enzymatic treatment [86].

5.1.2 Second generation biofuels

The production of second generation biofuels is still in progress. Here, bioethanol is produced from non-edible plant sources, from the biomass that is generally wasted.

This eliminates the food security concerns. Production of bioethanol from lignocelluloses involves pre-treatment of biomass to remove lignin, a constituent of the cell wall. The lignin presence in some plants hampers the activity of enzymes that degrade cellulose [87]. This cellulose is then broken down into simple sugars like glucose, which is then fermented to ethanol. In corn, for cellulase production, the catalytic domain of the β1,4-endoglucanase enzyme from *Acidothermus cellulolyticus* was fused with a Pr-1a signal peptide in tobacco plant. Later, this gene was introduced into the callus of maize plants using the gene gun method. Cellulase was seen to accumulate at 2.1% levels of total soluble proteins and could degrade cellulose [88].

A carbohydrate in wood named mannan can be degraded by mannanase. The β man gene isolated from the fungus, *Tricoderma reesei* was cloned and transferred to tobacco chloroplast. The tobacco leaf extracts were mixed with different mixture of enzymes like xylanase, esterase, endoglucanase, beta glucosidase. This experiment proved successful in hydrolyzing biomass. Therefore, the production of heterologous mannanase in plants is another way for cheap production of cellulosic bioethanol [89].

Many strategies have been implemented to produce bioethanol from plants and genetic engineering has played a vital role in increasing the bioethanol production to meet the ever-increasing demand of biofuel (**Figure 3**) [90].

5.2 Biopharming

Biopharming, molecular farming, or gene pharming is the development of transgenic plants and animals for the extraction of pharmaceutical substances that can be proteins, antibodies, vaccines, etc. for the benefit of humanity [91]. Plants utilized in biopharming are called as biological factories. The concept of molecular farming or "biopharming" was introduced by Fischer et al. to describe "the production of recombinant proteins in plants" and the products derived are called Plant-made pharmaceuticals (PMPs) [91, 92]. The process of developing PMPs includes identification the target protein, its respective gene, and the vectors that can deliver target DNA. Calli are developed from plant cells and then transferred into seed-producing plants, first in a greenhouse and then in fields. The desired protein is later purified from leaf or seed material [91]. The advantages of PMP over other established systems are low production costs, product safety, higher

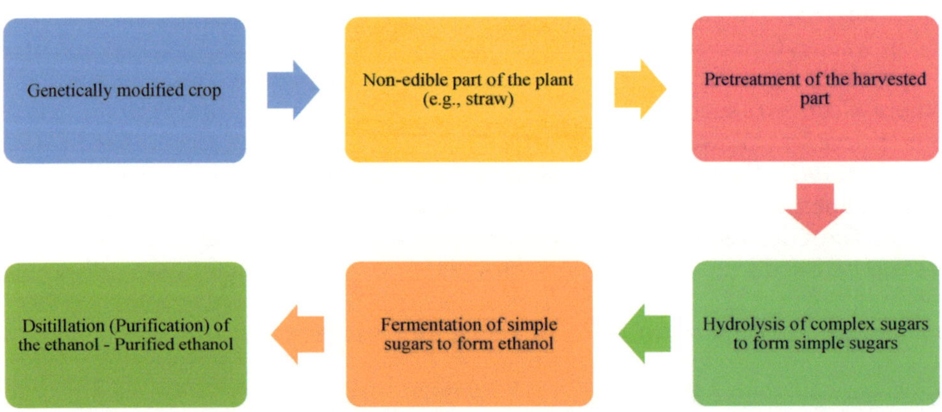

Figure 3.
Proposed pathway for production and purification of bioethanol from genetically modified crops (second generation biofuel).

scale up capacity, lower risk of contamination with human or animal pathogens, reduced downstream processing, and inherent capability of plant cells to perform complex post-translational modifications (**Table 5**) [91].

Prabavathy et al. [93] reviewed global concerns like inadequate crop production and malnutrition. Along with isolated nutrients, herbal products, dietary supplements, and nutraceuticals include genetically engineered designer foods. Genetic engineering techniques are applied to improve the nutritional and therapeutic value of the natural foods [93]. Habibi et al., seconded the use of transgenic roots as bio-factories to produce heterologous recombinant proteins for pharmaceutical applications [94]. Ahmad et al. [95] evaluated genetic modification to produce edible vaccines which are highly safe, cost-effective, and have many advantages over commercialized vaccines. Transgenic plants generated as edible vaccines express recombinant proteins including viral and bacterial antigens and antibodies e.g., banana, tomato, potato, spinach, tobacco, rice, carrot, corn. Edible vaccines are developed against disease like hepatitis B, cholera, HIV, measles, Norwalk virus, and rabies virus. The world most dangerous diseases like HIV and malaria can be effectively controlled using plant-based vaccines but since all the studies have been carried out to a limited scale, so for their effective widespread use, up-scaling of these studies are essential [95].

Joshi and Lopez [96] reviewed bioengineering in plants for therapeutic protein products like antibodies, vaccines, and plasma proteins while appreciating plants as an excellent system for the production of protein therapeutics alternative to microbial and mammalian reactor-based protein production technologies. Challenges such as bioequivalence, product consistency, challenges related to Post-Translational Protein Modifications (PTMs) are crucial to the structure and function of most eukaryotic proteins. One of such challenges is glycosylation. The 'humanization' of plant glycosylation pathways, combined with the discovery of terminal sialic acids (SAs) in plants making it feasible in plants to produce glycoproteins with mammalian-like

Crop	Disease against which biopharmaceuticals were produced
Lettuce	Malarial infection (CTB-MSP1 AMA-1), Dengue virus tetra-epitope peptide, hepatitis B
Tomato	Rabies
Soybean	Cholera infection
Cauliflower	Coronavirus pathogen in humans (proteins)
Rice	Porcine epidemic diarrhoea virus (PEDV) in pigs
Potato	Chicken infectious bronchitis virus vaccine
Collard	(Viral coat B5) Smallpox virus in humans
Peanut	Bovine rinderpest virus hemagglutinin (protein)
Arabidopsis	Foot and mouth disease viral structural VP1 (protein)
White clover	Bovine pneumonic pasteurellosis for rabbits (Leucotoxin antigen)
Alfalfa	Epitopes of bovine rotavirus VP4 protein for mice
Maize	(Spike protein antigen) porcine transmissible gastroenteritis virus (TGEV) for oral vaccine administration.
Barley	(Expression of F4 fimbriae) Enterotoxigenic *E. coli* causing diarrhoea in farm animals for piglets

Table 5.
Examples of PMPs developed for several diseases [91].

glycosylation. Molecular farming has been rapidly developed in plants as recombinant protein production systems. The therapeutic protein production is safe and immensely cost-effective in plants. Plant as a system is preferable over microbial fermentation and mammalian cell cultures due to their capability to perform post-translational modifications and to remain devoid of human infective viruses and prions. New plant expression systems have been recently developed to improve the yields and quality of plant-made pharmaceuticals, as well as to control post-translational maturations in transgenic plants, permitting for the production of human-like maturations in recombinant proteins [96, 97].

6. Biosensors to distinguish GM and non-GM foods

Sánchez-Paniagua et al. [98] described various biosensors based on distinct technology to distinguish between transgenic and non-transgenic crops or food and feed. The analytical methods for sensitive, accurate, rapid, and cheap detection of the products are required. DNA biosensors are conceived as a novel DNA-detection technology to substitute current amplification-based methods, providing hand-held, quick, and ultrasensitive gene-level detection [98]. Arugula et al. demonstrated real-time, label-free detection of DNA from genetically modified, Roundup Ready (RR) soybeans by surface plasmon resonance (SPR) in GM soybeans [99]. Huang et al. examined highly efficient analytical methods for rapid and high throughput screening of GMO components, as required for appropriate labeling of GMO-derived foods, as well as for on-site inspection and import/export quarantine [100].

7. Biosafety concerns: regulation of GM crops—a global perspective

Although Biotechnology contributes immensely to development in agricultural sectors, innovations and improvements in GM Crops have raised questions of their safety and effectiveness. As with any new technology, risks associated with it must be assessed and managed [101]. The definition of a GMO or GM crop given by the United Nations (UN) Cartagena Protocol on Biosafety is a "living modified organism" (LMO). A plant is genetically modified if satisfies two essential conditions, it contains a novel combination of genetic material, and the genetic material is introduced by using modern biotechnology. The GMO definition given in the Cartagena Protocol serves as the worldwide standard that individual countries can use to form their biosafety laws.

Recently, Site directed Nucleases (SDNs) have been used to create new varieties of crops. They can be of three types, SDN-1, SDN-2, and SDN-3. SDN-1 and SDN-2 can introduce random or specific mutations at specific loci, while SDN-3 target transgenes to a particular locus in the genome [102]. As opposed to traditional mutagenesis techniques used in plant breeding to generate variants, SDN mutagenesis requires the insertion of specific genes encoding the SDNs by transgenesis [103]. These genes are not essential for the maintenance of the crop and can be removed by segregation. Alternatively, SDNs can also be introduced without integrating the transgene into the host genome or by directly incorporating proteins [104]. With the usage of SDNs, new varieties of crops can be easily created with precision and predictability. Crops created using SDN-1 techniques are excluded from GM crop regulations in several countries like Germany, Netherlands, Argentina, Australia, Canada, Japan, and South Africa, as these crops introduce SDNs as proteins or RNA without usage of any transgenes [105].

In the US, the USDA has declared that SDN-1 crops will not fall under GM regulations as they are produced by incorporating the SDN transgene and subsequently removing it through segregation. For SDN 2 category crops, in some countries they do not fall under regulations like Germany in the EU. In case of SDN-3 category crops, they are considered as GMOs and fall under the purview of regulations [106].

In the case of GM crops, there is a difference between approvals for cultivation, for import and export, and consumption of GM crops as food and feed products [107]. This difference exists because of the different risks associated with cultivating, trading, and consuming, requiring different regulatory approaches. Multiple agencies are involved in granting of the approval request. In the United States, it is the responsibility of either the United States Department of Agriculture (USDA), the Environmental Protection Agency (EPA), the Food and Drug Administration (FDA), or more than one agency. GM regulations further can be classified as process or product-oriented. The focus is on the process used to produce the novel trait. The product-oriented regulations emphasize the novel characteristics of the product [106]. Canada remains the only country that has based their entire GM legislation on the product, rather than the process.

In the European Union (EU), Regulation (EC) No 1829/2003 on genetically modified food and feed binds all 27 Member States and specifically concerns GM food and feed produced "from" a GMO. This Regulation is applicable to food and feed products and their imports, in conjunction with Regulation 1830/2003 regarding tracking and labeling of GM products [108]. The European Network of GMO Laboratories (ENGL) has published a report on detecting food and feed products created by NBTs (New Breeding Techniques), identifying various possibilities and challenges [109]. EU relies on the GM regulation for products that are imported and so the obligation is on the developer of the gene-edited product to provide a suitable detection method for the product but no such products have been submitted for market authorization [110]. Since the EU imports most of their GM products, the regulations are more tailored to gene-edited food and feed products and not much on cultivation.

Norway and Switzerland ban the cultivation of GM crops in their national legislations. Norway does not cultivate GM crops nor does it import any GM food but GM crops are legally permitted by the Gene Technology Act [111]. The Norwegian Food Safety Authority has not yet approved any products or their deliberate release, except a single species of ornamental purple carnations [112, 113]. The Swiss legislation has suggested the usage of separate GM crop zones from 2021 for the cultivation of GM crops. This feature harbors the co-existence of GM crop cultivation with Non-GM Crop cultivation and would result in increase of acceptance of GM crops [114]. Similarly, Russia banned the cultivation of GM plants and breeding of GM animals under the amendments in Federal Law No. 358-FZ in July 2016 and with the recent approval of the new Food Security Doctrine in January 2020 [115].

The United States holds 30% of the whole market share and is a global leader in the commercialization of GM crops. Assessment of novel GM crop plant products can occur under three agencies, including the FDA, EPA, and USDA. The USDA's Animal and Plant Health Inspection Service (APHIS) ensures that introduction of GM plants does not pose a pest risk to plants. If the GM plant is intended for food use, the FDA evaluates the safety of the GM food product.

Canada is one the largest GM cultivators, producing 6% of the total global market share. As mentioned earlier, Canada uses product-based regulatory schemes. Here it is the mere presence of a novel trait, not the way it was introduced that is important [116, 117]. Irrespective of whether the characteristic was developed by new breeding techniques (NBTs) or mutagenesis or targeted mutagenesis, the GM crop product is subject to the same regulations by the Canadian Food Inspection Agency. FalcoTM Canola created by Cibus Canada Inc. (Cibus Canada Inc., 2020), is an herbicide-tolerant variety of canola. It was created by an oligonucleotide-directed mutagenesis (ODM) which caused a single nucleotide mutation in two genes. This ODM technique is very similar to CRISPR/Cas9 editing tool. The new canola variety was not considered different from the traditional canola variety (unmodified variety) by the Canadian Food Inspection Agency, giving it a non-GM crop status.

Commercial cultivation of GM crops is allowed in the following countries—India, China, Pakistan, Australia, Philippines, Myanmar, Vietnam, Bangladesh, and Indonesia [118]. India is both the world's largest Bt cotton producer. In January 2020, the Indian government, through its Department of Biotechnology, has published proposed gene editing guidelines. These guidelines suggest a multi-layered approach, with an increasing number of regulations for an increasing number of changes to the DNA. As SDN-1 encompasses genome editing using small insertions/deletions, and SDN-2 uses a small template for genome editing, plants edited using SDN-1 and SDN-2 are transgene free. Therefore, products of SDN-1 and SDN-2 (free from transgenes) are free from the provisions of Rules 7 and 11 (both inclusive) of the Rules, 1989. However, products of SDN-3 (with transgenes) come under regulation as applied to genetically engineered organisms under the Rules, 1989 [23].

Regulations on GM crop production, commercialization and trade differ throughout the globe. Nevertheless, concerns regarding GM crops remain valid and strict legislation requires rigorous scientific assessments in keeping up with societal values [102].

8. Conclusion

In the light of the ever-increasing and threatening repercussions of Global Climate Change, GMCs can be a promising driveway leading to sustainable approach toward addressing food crisis. Transgenic technologies can enhance crop yields and biotic and abiotic stress tolerance by imparting desirable genetic characteristics to crops. Transgenic crops can also be used as bio-factories for expression of high value recombinant proteins (biopharming) and bioethanol production. Agrobacterium-mediated and Biolistic methods by far remain the most applied tools in the development of GMCs. With the advent of genome editing technologies, a superior precision of genetic engineering can be achieved. But these protocols largely rely upon the existing transformation procedures for introduction of the genome editing vectors into the target tissue. Refining the parameters used for Agrobacterium-mediated and Biolistic transformation shall facilitate the development of optimized protocols for tissue culture and transformation thereby improving the transformation and genome editing efficiency. Biofortified crops appear to be promising candidates to address the global malnutrition issue. Exempting genome-edited products from regulation (especially in countries that suffer from malnutrition) shall facilitate the widespread distribution of such high value crops. Genome-edited crops have the potential to address climate change, food security, nutrition and health, environmental sustainability, and diversification of agriculture. But the concerns about the associated risks dampen the widespread enthusiasm about the future applications of genome-edited crops. The inability to address the regulatory, legal and trade framework, and social acceptance may hamper the potential benefits of genome-edited crops.

Thus, Genetically Modified Crops harbor the potential to increase the product yield, show improvement in the land use efficiency, and provide adequate nutrition for the ever-increasing population. This review has put a light on the importance, tools employed, applications, and safety of transgenic technology.

Acknowledgements

The authors are grateful to Dr. R.S. Zunjarrao, Principal, Modern College of Arts Science and Commerce (Autonomous) Shivajinagar, Pune 5, for the motivation and inspiration to be progressive in research field and Dr. Rama Phadke, HOD, Department of Biotechnology, Modern College of Arts Science and Commerce (Autonomous) Shivajinagar, Pune 5, for her guidance and support.

Conflict of interest

The authors declare that they have no conflict of interest.

Author details

Radhika Velankar*, Gauri Nerkar, Mukta Nagpurkar and Kiran Jagtap
Department of Biotechnology, Progressive Education Society's, Modern College of Arts, Science and Commerce (Autonomous), Shivajinagar, Pune, India

*Address all correspondence to: radhikaoke@gmail.com

References

[1] Bhatia S, Goli D. Introduction to Pharmaceutical Biotechnology, Basic Techniques and Concepts. Vol. 1. Philadelphia, USA: IOP Publishing; 2018

[2] Husby J. Definitions of GMO/LMO and modern biotechnology. In: Biosafety First: Holistic Approaches to Risk and Uncertainty in Genetic Engineering and Genetically Modified Organisms. Trondheim: Tapir Academic Press; 2007. pp. 569-581

[3] Lobato-Gómez M, Hewitt S, Capell T, Christou P, Dhingra A, Girón-Calva PS. Transgenic and genome-edited fruits: Background, constraints, benefits, and commercial opportunities. Horticulture Research. 2021;**8**:166. DOI: 10.1038/s41438-021-00601-3

[4] Trivedi M, Singh R, Shukla M, Tiwari RK. GMO and food security. In: Ecofriendly Pest Management for Food Security. USA: Academic Press; 2016. pp. 703-726

[5] Smyth SJ. Genetically modified crops, regulatory delays, and international trade. Food and Energy Security. 2017;**6**(2):78-86

[6] Lucht JM. Public acceptance of plant biotechnology and GM crops. Viruses. 2015;7(8):4254-4281

[7] Mustapa MAC, Batcha MFN, Amin L, Arham AF, Mahadi Z, Yusoff NMM, et al. Farmers' attitudes towards GM crops and their predictors. Journal of the Science of Food and Agriculture. 2021;**101**(13):5457-5468

[8] Zhaleh R, Mohammadi H, Boccia F, Firoozzare A, Covino D. Consumption of genetically modified food products, and its determinants (case study: Edible oil in Mashhad). Food. 2023;**12**(15):2933

[9] Rahnama H, Moradi AB, Moradi F, Noormohamadi N. Compositional and morphological analysis of salt stress tolerant Mannitol-1-phosphate dehydrogenase (mtlD)-transgenic potato plants. Plant Foods for Human Nutrition. 2023;**78**(4):670-675

[10] Bekele-Alemu A, Ligaba-Osena A. Comprehensive in silico analysis of the underutilized crop tef (Eragrostis tef (Zucc.) trotter) genome reveals drought tolerance signatures. BMC Plant Biology. 2023;**23**(1):506. DOI: 10.1186/s12870-023-04515-1

[11] Hansson SO. How to be cautious but open to learning: Time to update biotechnology and GMO legislation. Risk Analysis. 2016;**36**(8):1513-1517

[12] Waltz E. Non-browning GM apple cleared for market. Nature Biotechnology. 2015;**33**(4):326-328

[13] Gelvin SB. Integration of agrobacterium T-DNA into the plant genome. Annual Review of Genetics. 2017;**51**:195-217. DOI: 10.1146/annurev-genet-120215-035320

[14] Su W, Xu M, Radani Y, Yang L. Technological development and application of plant genetic transformation. International Journal of Molecular Science. 2023;**24**:10646. DOI: 10.3390/ijms241310646

[15] Klein T, Fromm M, Weissinger A, Tomes D, Schaaf S, Sletten M, et al. Transfer of foreign genes into intact maize cells with high velocity microprojectiles. Proceedings of the National Academy of Sciences, USA. 1988;**85**:4305-4309

[16] Christou P, McCabe DE, Swain WF. Stable transformation of soybean callus by DNA-coated gold particles. Plant

Physiology. 1988;**1988**(87):671-674. DOI: 10.1104/pp.87.3.671

[17] McCabe D, Martinell B. Transformation of elite cotton cultivars via particle bombardment of meristems. Nature Biotechnology. 1993;**11**:596-598. DOI: 10.1038/nbt0593-5966

[18] Anami S, Njuguna E, Coussens G, Aesaert S, Lijsebettens MV. Higher plant transformation: Principles and molecular tools. The International Journal of Developmental Biology. 2013;**57**:483-494. DOI: 10.1387/ijdb.130232mv

[19] Altpeter F, Baisakh N, Beachy R, Bock R, Capell T, Christou P, et al. Particle bombardment and the genetic enhancement of crops: Myths and realities. Molecular Breeding. 2005;**15**:305-327. DOI: 10.1007/s11032-004-8001-y

[20] Rao AQ, Bakhsh A, Kiani S, Shahzad K, Shahid AA, Husnain T, et al. The myth of plant transformation. Biotechnology Advances. 2009;**27**:753-763. DOI: 10.1016/j.biotechadv.2019.06.003

[21] Christou P. Transformation technology. Trends in Plant Science. 1996;**1**(12):423-431. DOI: 10.1016/S1360-1385(96)10047-9

[22] Keshavareddy G, Kumar ARV, Ramu VS. Methods of plant transformation—A review. International Journal of Current Microbiology and Applied Sciences. 2018;**7**:2656-2668. DOI: 10.20546/ijcmas.2018.707.312

[23] Nerkar G, Devarumath S, Purankar M, Kumar A, Valarmathi R, Devarumath R, et al. Advances in crop breeding through precision genome editing. Frontiers in Genetics. 2022;**13**:880195. DOI: 10.3389/fgene.2022.880195

[24] Puchta H. The repair of double-strand breaks in plants: Mechanisms and consequences for genome evolution. Journal of Experimental Botany. 2005;**56**:1-14. DOI: 10.1093/jxb/eri025

[25] Sun Y, Li J, Xia L. Precise genome modification via sequence-specific nucleases-mediated gene targeting for crop improvement. Frontiers in Plant Science. 2016;**7**:1928. DOI: 10.3389/fpls.2016.01928

[26] Symington LS, Gautier J. Double-strand break end resection and repair pathway choice. Annual Review of Genetics. 2011;**45**(1):247-271. DOI: 10.1146/annurev-genet-110410-132435

[27] Smith JS, Grizot S, Arnould, et al. A combinatorial approach to create artificial homing endonucleases cleaving chosen sequences. Nucleic Acids Research. 2006;**34**(22):e149. DOI: 10.1093/nar/gkl720

[28] Kim YG, Cha J, Chandrasegaran S. Hybrid restriction enzymes: Zinc finger fusions to FokI cleavage domain. Proceedings of National Academy of Sciences USA. 1996;**93**:1156-1160. DOI: 10.1073/pnas.93.3.1156

[29] Miller J, McLachlan AD, Klug A. Repetitive zinc-binding domains in the protein transcription factor IIIA from xenopus oocytes. The Journal of Trace Elements in Experimental Medicine. 2001;**14**(2):157-169

[30] Bae KH, Kwon YD, Shin HC. Human zinc fingers as building blocks in the construction of artificial transcription factors. Nature Biotechnology. 2003;**21**:275-280

[31] Christian MT, Cermak EL, Doyle C, Schmidt F, Zhang A, Hummel AJ, et al. Targeting DNA double-strand breaks with TAL effector nucleases. Genetics. 2010;**186**(2):757-761. DOI: 10.1534/genetics.110.120717

[32] Deng D, Yan C, Pan X, et al. Structural basis for sequence specific recognition of DNA by TAL effectors. Science. 2012;**335**(6069):720-723

[33] Boch J, Scholze H, Schornack S, et al. Breaking the code of DNA binding specificity of TAL-type III effectors. Science. 2009;**326**(5959):1509-1512

[34] Jinek M, Chylinski K, Fonfara I, Hauer M, Doudna JA, Charpentier E. A programmable dual-RNA-guided DNA endonuclease in adaptive bacterial immunity. Science. 2012;**337**(6096):816-821. DOI: 10.1126/science.1225829

[35] Basso MF, Arraes FBM, Grossi-de-Sa M, Moreira VJV, Alves-Ferreira M, Grossi-de-Sa MF. Insights into genetic and molecular elements for transgenic crop development. Frontiers in Plant Science. 2020;**11**:509. DOI: 10.3389/fpls.2020.00509

[36] Nawaz M, Sun J, Shabbir S, Khattak WA, Ren G, Nie X, et al. A review of plants strategies to resist biotic and abiotic environmental stressors. Science of Total Environment. 2023;**900**:165832. DOI: 10.1016/scitotenv.2023.165832

[37] Esmaeili N, Shen G, Zhang H. Genetic manipulation for abiotic stress resistance traits in crops. Frontiers in Plant Science. 2022;**13**:1011985. DOI: 10.3389/fpls.2022.1011985

[38] Chinnusamy V, Zhu J, Zhu JK. Cold stress regulation of gene expression in plants. Trends in Plant Science. 2007;**12**(10):444-451. DOI: 10.1016/j.tplants.2007.07.002

[39] Zandalinas SI, Fritschi FB, Mittler R. Global warming, climate change, and environmental pollution: Recipe for a multifactorial stress combination disaster. Trends in Plant Science. 2021;**26**(6):588-599. DOI: 10.1016/j.tplants.2021.02.011

[40] Sinha R, Irulappan V, Mohan-Raju B, et al. Impact of drought stress on simultaneously occurring pathogen infection in field-grown chickpea. Nature Scientific Reports. 2019;**9**:5577. DOI: 10.1038/s41598-019-41463-z

[41] Rivero RM, Mittler R, Blumwald E, Zandalinas SI. Developing climate resilient crops: Improving plant tolerance to stress combination. The Plant Journal. 2022;**109**:373-389. DOI: 10.1111/tpj.15483

[42] ISAAA. Global Status of Commercialized Biotech/GM Crops in 2019: Biotech Crops Drive SocioEconomic Development and Sustainable Environment in the New Frontier. The International Service for the Acquisition of Agri-Biotech Applications (ISAAA). Brief No. 55. ISAAA: Ithaca, NY; 2019

[43] Razzaq M, Akhter M, Ahmad R, Cheema K, Hina A, Karikari B, et al. CRISPR-Cas9 based stress tolerance: New hope for abiotic stress tolerance in chickpea (*Cicer arietinum*). Molecular Biology Reports. 2022;**2022**:49. DOI: 10.1007/s11033-022-07391-4

[44] Solanke AU, Arora K, Karkute SG, Tomar RSS. Genetic engineering for biotic stress management in rice. In: Sarmah BK, Borah BK, editors. Genome Engineering for Crop Improvement. Concepts and Strategies in Plant Sciences. Cham: Springer; 2021. pp. 117-139. DOI: 10.1007/978-3-030-63372-1_5

[45] Hmida-Sayari A, Gargouri-Bouzid R, Bidani A, Jaoua L, Savouré A, Jaoua S. Overexpression of Δ1-pyrroline-5-carboxylate synthetase increases proline production and confers salt tolerance in transgenic potato plants. Plant Science. 2005;**169**(4):746-752. DOI: 10.1016/j.plantsci.2005.05.025

[46] Altpeter F, Oraby H. Sugarcane. In: Kempken F, Jung C, editors. Biotechnology in Agriculture and Forestry. New York: Springer; 2010. pp. 453-472. DOI: 10.1007/978-3-642-02391-0

[47] Hotta CT, Lembke CG, Domingues DS, Ochoa EA, Cruz GMQ, Melotto-Passarin DM, et al. The biotechnology roadmap for sugarcane improvement. Tropical Plant Biology. 2011;**3**(2):75-87. DOI: 10.1007/s12042-010-9050-5

[48] Kumar T, Uzma K, M.R., et al. Genetic improvement of sugarcane for drought and salinity stress tolerance using Arabidopsis vacuolar pyrophosphatase (AVP1) gene. Molcular Biotechnology. 2014;**56**:199-209. DOI: 10.1007/s12033-013-9695-z

[49] Purayannur S, Kumar K, Verma PK. Genetic engineering to improve biotic stress tolerance in plants. In: Abdin M, Kiran U, Kamaluddin A, editors. Plant Biotechnology: Principles and Applications. Singapore: Springer; 2017. pp. 207-232. DOI: 10.1007/978-981-10-2961-5_8

[50] Kumar K, Gambhir G, Dass A, Tripathi AK, Singh A, Jha AK, et al. Genetically modified crops: Current status, and future prospects. Planta. 2020;**251**(4):1-27. DOI: 10.1007/s00425-020-03372-8

[51] Savadi S, Lambani N, Kashyap P, Bisht D. Genetic engineering approaches to enhance oil content in oilseed crops. Plant Growth Regulation. 2017;**83**. DOI: 10.1007/s10725-016-0236-1

[52] Liu WX, Liu HL, Qu LQ. Embryo-specific expression of soy bean oleosin altered oil body morphogenesis and increased lipid content in transgenic rice seeds. Theoretical and Applied Genetics. 2013;**126**:2289-2297

[53] Zhang H, Zhang J, Wei P, Zhang B, Gou F, Feng Z, et al. The CRISPR/Cas9 system produces specific and homozygous targeted gene editing in rice in one generation. Plant Biotechnology Journal. 2014;**12**:797-807

[54] Chen L, Hao L, Parry MA, Phillips AL, Hu YG. Progress in TILLING as a tool for functional genomics and improvement of crops. Journal of Integrative Plant Biology. 2014;**56**:425-443

[55] Petrie JR, Vanhercke T, Shrestha P, El-Tahchy A, White A, Zhou XR, et al. Recruiting a new substrate for triacylglycerol synthesis in plants: The monoacylglycerolacyltransferase pathway. PLoS One. 2012;7:c35214

[56] Vanhercke T, El Tahchy A, Liu Q, Zhou XR, Shrestha P, Divi UK, et al. Metabolic engineering of biomass for high energy density: Oilseed-like triacylglycerol yields from plant leaves. Plant Biotechnology Journal. 2014;**12**(2):231-239. DOI: 10.1111/pbi.12131

[57] Garg M, Sharma N, Sharma S, Kapoor P, Kumar A, Chunduri V, et al. Biofortified crops generated by breeding, agronomy, and transgenic approaches are improving lives of millions of people around the world. Frontiers in Nutrition. 2018;**5**:12. DOI: 10.3389/fnut.2018.00012

[58] Burkhardt PK, Beyer P, Wuenn J, Kloeti A, Armstrong GA, Schledz M, et al. Transgenic rice (*Oryza sativa*) endosperm expressing daffodil (*Narcissus pseudonarcissus*) phytoene synthase accumulates phytoene, a key intermediate of provitamin A biosynthesis. Plant Journal. 1997;**11**:1071-1078. DOI: 10.1046/j.1365-313X.1997.11051071.x

[59] Blancquaert D, Van Daele J, Strobbe S, Kiekens F, Storozhenko S, De Steur H, et al. Improving folate (vitamin B9) stability in biofortified rice through metabolic engineering. Nature Biotechnology. 2015;**33**:1076-1078. DOI: 10.1038/nbt.3358

[60] Cong L, Wang C, Chen L, Liu H, Yang G, He G. Expression of phytoene

synthase1 and carotene desaturase crtI genes result in an increase in the total carotenoids content in transgenic elite wheat (*Triticum aestivum* L.). Journal of Agricultural and Food Chemistry. 2009;**57**(18):8652-8660. DOI: 10.1021/jf9012218

[61] Xiaoyan S, Yan Z, Shubin W. Improvement Fe content of wheat (*Triticum aestivum*) grain by soybean ferritin expression cassette without vector backbone sequence. Agricultural Biotechnology Journal. 2012;**20**:766-773

[62] Tamas C, Kisgyorgy BN, Rakszegi M, Wilkinson MD, Yang MS, Lang L, et al. Transgenic approach to improve wheat (*Triticum aestivum* L.) nutritional quality. Plant Cell Reports. 2009;**28**(7):1085-1094. DOI: 10.1007/s00299-009-0716-0

[63] Doshi KM, Eudes F, Laroche A, Gaudet D. Transient embryo specific expression of anthocyanin in wheat. In Vitro Cellular and Developmental Biology-Plant. 2006;**42**:432-438. DOI: 10.1079/IVP2006778

[64] Sestili F, Janni M, Doherty A, Botticella E, D'Ovidio R, Masci S, et al. Increasing the amylose content of durum wheat through silencing of the SBEIIa genes. BMC Plant Biology. 2010;**10**:144. DOI: 10.1186/1471-2229-10-144

[65] Ramesh SA, Choimes S, Schachtman DP. Over-expression of an Arabidopsis zinc transporter in Hordeum vulgare increases short-term zinc uptake after zinc deprivation and seed zinc content. Plant Molecular Biology. 2004;**54**(3):373-385. DOI: 10.1023/B:PLAN.0000036370.70912.34

[66] Elkonin LA, ItalianskayaI JV, Domanina VN, Selivanov NY, Rakitin AL, Ravi NV. Transgenic sorghum with improved digestibility of storage proteins obtained by agrobacterium-mediated transformation. Russian Journal of Plant Physiology. 2016;**63**:678-689. DOI: 10.1134/S1021443716050046

[67] Schmidt MA, Parrott WA, Hildebrand DF, Berg RH, Cooksey A, Pendarvis K, et al. Transgenic soya bean seeds accumulating β-carotene exhibit the collateral enhancements of oleate and protein content traits. Plant Biotechnology Journal. 2015;**13**(4):590-600. DOI: 10.1111/pbi.12286

[68] Diretto G, Tavazza R, Welsch R, Pizzichini D, Mourgues F, Papacchioli V, et al. Metabolic engineering of potato tuber carotenoids through tuber-specific silencing of lycopene epsilon cyclase. BMC Plant Biology. 2006;**6**:13. DOI: 10.1186/1471-2229-6-13

[69] Morris J, Hawthorne KM, Hotze T, Abrams SA, Hirschi KD. Nutritional impact of elevated calcium transporter activity in carrots. Proceedings of the National Academy of Sciences of the United States of America. 2008;**105**:1431-1435. DOI: 10.1073/pnas.0709005105

[70] Fraser PD, Enfissi EM, Halket JM, Truesdale MR, Yu D, Gerrish C, et al. Manipulation of phytoene levels in tomato fruit: Effects on isoprenoids, plastids, and intermediary metabolism. The Plant Cell. 2007;**19**(10):3194-3211. DOI: 10.1105/tpc.106.049817

[71] Waltz E. Vitamin A super banana in human trials. Nature Biotechnology. 2014;**32**:857. DOI: 10.1038/nbt0914-857

[72] Avaram T, Badani H, Galili S, Aamir R. Enhanced levels of methionine and cysteine in transgenic alfalfa (*Medicago sativa* L.) plants over-expressing the Arabidopsis cystathionine γ-synthase gene. Plant Biotechnology Journal. 2005;**3**:71-79. DOI: 10.1111/j.1467-7652.2004.00102.x

[73] Wang D, Ziska LH, Xu X, Tao Y, Zhang J, Liu G, et al. Adapting rice to rising atmospheric carbon dioxide: A preliminary GMO approach to maintain nutritional integrity. European Journal of Agronomy. 2023;**144**:126766

[74] Kumar V, Sharma N, Jaiswal KK, Vlaskin MS, Nanda M, Tripathi MK, et al. Microalgae with a truncated light-harvesting antenna to maximize photosynthetic efficiency and biomass productivity: Recent advances and current challenges. Process Biochemistry. 2021;**104**:83-91

[75] Éva C, Oszvald M, Tamás L. Current and possible approaches for improving photosynthetic efficiency. Plant Science: An International Journal of Experimental Plant Biology. 2019;**280**:433-440. DOI: 10.1016/j.plantsci.2018.11.010

[76] Brookes G. Environmental impacts of genetically modified (GM) crop use: Impacts on pesticide use and carbon emissions. In: Plant Biotechnology: Experience and Future Prospects. USA: Springer International Publishing; 2021. pp. 87-101

[77] Sutherland C, Gleim S, Smyth SJ. Correlating genetically modified crops, glyphosate use and increased carbon sequestration. Sustainability. 2021;**13**(21):11679

[78] Van Acker R, Rahman MM, Cici SZH. Pros and cons of GMO crop farming. In: Oxford Research Encyclopedia of Environmental Science. USA; 2017. DOI: 10.3390/v7082819

[79] Rubin EM. Genomics of cellulosic biofuels. Nature. 2008;**454**:841-845

[80] Gressel J. Transgenics are imperative for biofuel crops. Plant Science. 2008;**174**(3):246-263. DOI: 10.1016/j.plantsci.2007.11.009

[81] Lee D, Chen A, Nair R. Genetically engineered crops for biofuel production: Regulatory perspectives. Biotechnology & Genetic Engineering Reviews. 2008;**2008**(25):331-361. DOI: 10.5661/bger-25-331

[82] Bordetsky A, Hwang R, Korin A, Lovaas D, Tonachel L. Securing America: Solving Our Oil Dependence through Innovation. New York: Natural Resources Defence Council; 2005

[83] Saha S, Ramachandran S. Genetic improvement of plants for enhanced bio-ethanol production. Recent Patents on DNA & Gene Sequences. 2013;**7**(1):36-44. DOI: 10.2174/1872215611307010006

[84] Wu L, Birch RG. Doubled sugar content in sugarcane plants modified to produce a sucrose isomer. Plant Biotechnology Journal. 2007;**5**(1):109-117

[85] Martin AM. First generation biofuels compete. New Biotechnology. 2010;**27**(5):596-608

[86] Torney F, Moeller L, Scarpa A, Wang K. Genetic engineering approaches to improve bioethanol production from maize. Current Opinion in Biotechnology. 2007;**18**(3):193-199

[87] Cook C, Devoto A. Fuel from plant cell walls: Recent developments in second generation bioethanol research. Journal of the Science of Food and Agriculture. 2011;**91**(10):1729-1732. DOI: 10.1002/jsfa.4455

[88] Biswas GCG, Ransom C. Expression of biologically active Acidothermus cellulolyticus endoglucanase in transgenic maize plants. Plant Science. 2006;**171**(5):617-623. DOI: 10.1016/j.plantsci.2006.06.004

[89] Agarwal P, Verma D, Daniell H. Expression of Trichoderma

reesei-Mannanase in tobacco chloroplasts and its utilization in lignocellulosic woody biomass hydrolysis. PLoS One. 2011;**6**(12):1-10

[90] Rocha-Meneses L, Ferreira JA, Mushtaq M, Karimi S, Orupõld K, Kikas T. Genetic modification of cereal plants: A strategy to enhance bioethanol yields from agricultural waste. Industrial Crops and Products. 2020;**150**:112408. DOI: 10.1016/j.indcrop.2020.112408

[91] Datta M, Satapathy S, Kumar S, Vir R, Kumar V. An encounter with biopharming. In: 4th Global Meet on Science and Technology. Vol. 44. Uttar Pradesh, India: Global Meet; 2021

[92] Fischer R, Drossard J, Commandeur U, Schillberg S, Emans N. Towards molecular farming in the future: Moving from diagnostic protein and antibody production in microbes to plants. Biotechnology and Applied Biochemistry. 1999;**30**:101-108

[93] Prabavathy D, Sudha S, Inbathamizh L. Genetically modified products and non-GMO products in nutraceuticals. Handbook of Nutraceuticals and Natural Products: Biological, Medicinal, and Nutritional Properties and Applications. 2022;**2**:315-336

[94] Habibi P, De Sa MFG, Makhzoum A, Malik S, da Silva ALL, Hefferon K, et al. Bioengineering hairy roots: Phytoremediation, secondary metabolism, molecular pharming, plant-plant interactions, and biofuels. Sustainable Agriculture Reviews. 2017;**22**:213-251

[95] Ahmad P, Ashraf M, Younis M, Hu X, Kumar A, Akram NA, et al. Role of transgenic plants in agriculture and biopharming. Biotechnology Advances. 2012;**30**(3):524-540. DOI: 10.1016/j.biotechadv.2011.09.006

[96] Joshi L, Lopez LC. Bioprospecting in plants for engineered proteins. Current Opinion in Plant Biology. 2005;**8**(2):223-226

[97] Lienard D, Sourrouille C, Gomord V, Faye L. Pharming, and transgenic plants. Biotechnology Annual Review. 2007;**13**:115-147

[98] Sánchez-Paniagua López M, Manzanares-Palenzuela CL, López-Ruiz B. Biosensors for GMO testing: Nearly 25 years of research. Critical Reviews in Analytical Chemistry. 2018;**48**(5):391-405

[99] Arugula M, Chanysheva A, Vaglenov K, Simonian A. Biosensors for detecting genetically modified organisms in food and feed. ECS Transactions. 2015;**66**(36):31. DOI: 10.1149/06636.0031ecst

[100] Huang L, Zheng L, Chen Y, Xue F, Cheng L, Adeloju SB, et al. A novel GMO biosensor for rapid ultrasensitive and simultaneous detection of multiple DNA components in GMO products. Biosensors and Bioelectronics. 2015;**66**:431-437

[101] Aven T. Risk assessment and risk management: Review of recent advances on their foundation. European Journal of Operational Research. 2016;**253**(1):1-13. Available from: https://www.sciencedirect.com/science/article/pii/S0377221715011479

[102] Podevin N, Davies HV, Hartung F, Casacuberta JM, Nogue F. Site-directed nucleases: A paradigm shift in predictable, knowledge-based plant breeding. Trends in Biotechnology. 2013;**31**(6):375-383

[103] Tzfira T, Weinthal D, Marton I, Zeevi V, Zuker A, Vainstein A. Genome modifications in plant

cells by custom-made restriction enzymes. Plant Biotechnology Journal. 2012;**10**(4):373-389

[104] Gaj T, Guo J, Kato Y, Sirk SJ, Barbas CF, Waltz E. Tiptoeing around transgenics. Nature Methods. 2012;**9**(8):805-807. DOI: 10.1038/nbt.2143

[105] Lusser M, Davies HV. Comparative regulatory approaches for new plant breeding techniques. In: Workshop Proceedings. Luxembourg (Luxembourg): Publications Office of the European Union; 2012. Available from: http://ipts.jrc.ec.europa.eu/publications/pub.cfm?id=4959

[106] McHughen A. A critical assessment of regulatory triggers for products of biotechnology: Product vs. process. GM Crops Food. 2016;**7**(3-4):125-158

[107] Eckerstorfer MF, Engelhard M, Heissenberger A, Simon S, Teichmann H. Plants developed by new genetic modification techniques-comparison of existing regulatory frameworks in the EU and non-EU countries. Frontiers in Bioengineering and Biotechnology. 2019;**7**(26):1-16. DOI: 10.3389/fbioe.2019.00026

[108] Hundleby PAC, Harwood WA. Impacts of the EU GMO regulatory framework for plant genome editing. Food and Energy Security. 2019;**8**(2):1-8

[109] ENGL. Detection of food and feed plant products obtained by new mutagenesis techniques. 2019. Available from: https://gmo-crl.jrc.ec.europa.eu/doc/JRC116289-GE-report-ENGL.pdf

[110] Lombardo L, Grando MS. Genetically modified plants for nutritionally improved food: A promise kept? Food Reviews International. 2020;**36**(1):58-76. DOI: 10.1080/87559129.2019.1613664

[111] Bratlie S, Halvorsen K, Myskja BK, Mellegård H, Bjorvatn C, Frost P, et al. A novel governance framework for GMO. EMBO Reports. 2019;**20**(5):e47812, 10.15252/embr.201947812

[112] Mattilsynet. Matportalen: Genmodifisert Mat (GMO). 2020. Available from: https://www.matportalen.no/merking/tema/genmodifisert_mat_gmo/

[113] ISAAA GM Approval Database. 2020. Available from: http://www.isaaa.org/gmapprovaldatabase/default.asp [Accessed: October 14, 2020]

[114] Chandrasekhar A. Government approves GMO ban extension. SWI. 2016 Available from: https://www.swissinfo.ch/eng/society/genetically-modified-organisms_government-approves-gmo-ban-extension/42260828

[115] Galata BE. The framing effect of the media in the regulation of GMOs: A case study of Russia. Russian Journal of Communication. 2019;**11**(3):240-252. DOI: 10.1080/19409419.2019.1678348

[116] Whelan AI, Lema MA. Regulatory framework for gene editing and other new breeding techniques (NBTs) in Argentina. GM Crops Food. 2015;**6**(4):253-265

[117] CFIA. Plants with novel traits. 2020. Available from: https://www.inspection.gc.ca/plant-varieties/plants-with-novel-traits/eng/1300137887237/1300137939635

[118] ISAAA. ISAAA Brief 55-2019: Executive Summary. 2020. Available from: https://www.isaaa.org/resources/publications/briefs/55/executivesummary/

Chapter 3

Mutagenesis Application in Plant Improvement: Advancements and Its Future

Augustine Antwi-Boasiako, Padmore Adu-Antwi, Richard Adu Amoah, Augustine Boakye Boateng, Joseph Gyau, Matilda Frimpong, Isaac Newton Boakye-Mensah and Ivy Odi Ahiamadia

Abstract

Agricultural plant genetic resources are constantly going into extinction having negative implications for plant genetic banks. Hence, there is a need to generate variations. Stimulated mutagenesis offers an efficient tool to generate genetic variation and explore the function of genes. It also facilitates the identification of genes and their roles in traits of economic interest to breeders, farmers and consumers. Thus, transforming the agro-based industries in overcoming obstacles (poor yield, lodging, shattering, pests and disease infestations). Exploring alternatives to integrate farmers' and consumers' desirable traits into their preferred cultivars has led to major advancements in mutation breeding. The chapter provides a comprehensive update on induced mutagenesis approaches, increasing efficiency of targeted mutagenesis and identification of novel traits in mutated populations. Furthermore, it reveals the efforts of ten countries that are leading the development of varieties via mutation across the globe and the most prioritised crops that have received critical attention in mutation breeding. Moreover, it seeks to bring to light the current approaches used in facilitating mutation breeding. It details the current progress made in improving plants with evidence relating to generating genetic resources, biotic and abiotic stresses, nutritional, and quality improvement whiles providing future directions for mutation breeding.

Keywords: mutagenesis, mutation breeding, plant improvement, plant biodiversity, genetic resources, stress tolerance, trait improvement

1. Introduction

Genetic variation is pivotal to ensuring the evolutionary adaptation of plant species. It also facilitates plants' adaptation to various stresses arising from their environment. Many interventions are designed to promote increased genetic diversity in plants not excluding mutation [1]. Mutation occurs when there is a sudden change in the sequences of the DNA but it is not derived through segregation or recombination. Mutagenesis is the process by which mutation occurs. Plant mutation breeding also termed variation breeding employs the use of physical radiations, chemical means and/or biological mutagens to cause spontaneous genetics. Several methods are used to induce mutagenesis in plants [2]. Physical mutagens (UV, X-ray, fast neutron, as well as gamma radiation); chemical mutagens such as EMS (ethyl methanesulfonate), MNU (N-methyl-N-nitrosourea), HF (hydrogen fluoride), MMS (methyl methane-sulfonate) and biological mutagens (Agrobacterium and transposon-based chromosomal integration) are broadly explored. These methods are quite tedious.

The advent of molecular genetic techniques has developed the acquisition and study of mutations. For instance, in vitro mutagenesis protocols are been established in plants (wheat, saffron, and dendrobium) to achieve maximum variability for subsequent characterisation by investigating the germplasm at various levels including molecular, biochemical, and morphology [3–5]. In vitro mutagenesis has successfully been employed in many plant species to overcome abiotic (drought, salt, frost, and aluminium) and biotic stresses [6]. In addition, next-generation sequencing (NGS) detects mutations very quickly, and it is cost-effective. Integrating induced mutation with whole-genome sequencing supports the use of forward and reverse genetics enabling specific genome editing techniques such as ZFNs (Zinc Finger Nucleases), transcription activator-like effector nucleases (TALENS), and clustered regularly interspaced short palindromic repeats (CRISPR)/CRISPR-associated9 (Cas9) endonuclease. The mutants generated are termed mutagenic plants or seeds. They become potential commercial varieties or are used as a parent plant to generate new varieties.

Mutation breeding serves as an avenue for generating demand-driven varieties that meet the needs of customers, industries, and producers. This is made possible by supporting the functional characterisation of genes, and the establishment of gene-trait relationships among numerous crop types. Hence, mutation drives the drift and leading methods in evolution and plant breeding [2, 7]. This account for the continuous use of mutation breeding by breeders after its discovery in 1925 as they are considered as not transgenic in nations where transgenic plants are denied [8]. The advancement in mutation breeding has been discussed to reveal its relevance in the crop improvement process and its future directions. This chapter offers a comprehensive overview of the progresses induced mutagenesis provides in achieving zero hunger by contributing to food security in the light of climate change. Specifically, the chapter presents i) mutagenesis approaches for plant improvement ii) increasing efficiency of targeted mutagenesis and identification of some novel traits iii) contributions of mutagenesis in plant improvement (plant diversity & genetic resources; biotic & abiotic stress resistance; & nutritional quality trait improvement) and iv) future of mutagenesis for improving plants.

2. Mutagenesis approaches for plant improvement

Mutation breeding techniques have extensively been employed for crop enhancement. The techniques involve the use of both physical, chemical, biological mutagens, and gene editing (CRISPR/Cas9, TALEN-based, VIGS, RNAi interference) to generate genetic variability in crops and their effects on crops have been well characterised [2, 9–11]. The physical mutagens commonly used include gamma rays, alpha particles, X-rays, fast neutrons, beta particles and UV light. Also, ethyl methanesulfonate (EMS), N-methyl-N-nitrosourea (MNU), sodium azides, diethyl

DOI: http://dx.doi.org/10.5772/intechopen.112510

Mutagen	Crops	Gene(s)	Mutation type	Pathogen/ weed type	Disease/weed/ Herbicide name	Evaluation environment	Factors effect	Reference
GR	Rice (*O. sativa*)	*Pi*		Fungus	Blast		Improves resistance	[16]
GR	Rice (*O. sativa*)			Fungus	Blast	Greenhouse	Induces resistance	[17]
GR	Potato (*Solanum tuberosum L.*)			Fungus	Stem canker and black scurf	Laboratory	Improves resistance	[18]
GR	Sesame (*Sesamum indicum L.*)			Fugus	Phytophthora blight	Laboratory & field	Enhances resistance to phytophthora blight	[19]
FNI	Rice (*O. sativa*)	*Snl6*	Base deletion	Bacteria	Bacterial blight	Field	Improves resistance	[20]
GR & EMS	Chickpea (*Cicer arietinum*)			Fungus	Fusarium wilt	Field	resistance to Fusarium wilt	[21]
GR	Tomato (*Solanum lycopersicum*)			Bacteria	Bacterial wilt resistance	Laboratory & field	Enhances resistance	[22]
GR	Rice (*O. sativa*)			Fungus	Blast	Field	Provides broad-spectrum resistance	[23]
GR	Sugar cane (*Saccharum* species)		Base deletion	Virus	Sugarcane mosaic virus	Greenhouse & field	Tolerance to sugarcane Mosaic Virus (SCMV)	[24]
UV	Arabidopsis (*A. thaliana*)			Fungus	Downy mildew	Growth chamber	Promote resistance	[25]
UV	Tea (*Camellia sinensis*)			Fungus	Blister blight	Field	Tolerance to blister blight	[26]
EMS	Soybean (*Glycine max*)	*AHAS*	Base substitution	Herbicides	Sulfonylureas herbicide	Field & greenhouse	Chlorsulfuron-Resistant Soybean	[27]
EMS	Rice (*O. sativa*)			Insect & Bacteria	yellow stem borer, sheath blight & bacterial leaf blight	Field & Laboratory	Enhance rice tolerance to insects and bacteria	[28]
NaN3	Rice (O. sativa)			Fungus & Bacteria	Blast & bacterial blight		Provides broad-spectrumresistance	[29]
EMS	Sesame (*S. indicum*)			Fungus	Phytophthora blight	Laboratory & Field	Resistance to phytophthora blight	[19]

Mutagen	Crops	Gene(s)	Mutation type	Pathogen/ weed type	Disease/weed/ Herbicide name	Evaluation environment	Factors effect	Reference
EMS	Sweet orange (*Citrus sinensis*)			Bacteria	citrus canker	In-vitro & in-vivo	Tolerant to Citrus canker	[30]
NaN3	Sorghum (*Sorghum bicolor* L.), rice (*O. sativa*), barley (*Hordeum vulgare*), maize (*Zea mays*)			Parasitic weed	*Striga hermonthica*	Field	Suppress the growth of striga in maize	[31]
CRISPR/ Cas9	Rice (*O. sativa*)	*OsHPP04*	Base deletion	Nematode	root-knot nematode		Enhances resistance	[32]
CRISPR/ Cas9	Soybean (*G. max*)	*GmTCP19L*	Base deletion	Fungus	Phytophthora root rot	Sequencing analysis	Increase susceptibility	[33]
CRISPR/ Cas9	Tomato (*S. lycopersicum*)	*SlDMR6–1* & *SlDMR6–2*	Base deletion	Bacteria & fungus	Bacterial spot & powdery mildew	Field	Enhance resistance to bacterial, and fungal pathogens	[34]
CRISPR/ Cas9	Rice (*O. sativa* L.)	*OsAvrXa7*	Base deletion	Bacteria	Bacterial blight	Greenhouse	Bacterial blight resistance	[35]
VIGS & CRISPR	Wheat (*Triticum aestivum* L.)	*TaNFXL1*	Base insertion/ Deletion	Fungus	Fusarium head blight	Growth chamber	Promotes resistance	[11]
CRISPR/ Cas9	Tomato (*S. lycopersicum*)	*SlPMR4*	Base insertion/ deletion/ inversion	Fungus	Powdery mildew	Greenhouse	Enhances resistance	[36]
CRISPR/ Cas9	Tomato (*S. lycopersicum*)	*SleIF4E1*	Base deletion	Virus	Mottle virus	Growth chamber	Confers resistance	[37]
CRISPR/ Cas9	Tomato (*Solanum lycopersicum* L.)	*PMR4*	Base deletion	Fungus	Powdery mildew		Promotes resistances	[36]
CRISPR/ Cas9	Grapevine (*Vitis vinifera*)	*VvMLO3*	Base insertion/ deletion	Fungus	Powdery mildew	Laboratory	Enhances resistance	[38]

Mutagen	Crops	Gene(s)	Mutation type	Pathogen/ weed type	Disease/weed/ Herbicide name	Evaluation environment	Factors effect	Reference
CRISPR/ Cas9	rice (*O. sativa*)	*OsSWEET11, 13 & 14*	Base insertion/ deletion	Bacteria	Bacterial blight		Bacterial blight resistance	[39]
CRISPR/ Cas9	rice (*O. sativa*)	OsXa13/Os8N3	Base deletion	Bacteria	Bacterial blight	Greenhouse	Enhance resistance	[40]
CRISPR/ Cas9	Tomato (*S. lycopersicum*)	*SlJAZ2*	Base deletion	Bacteria	Bacterial speck	Growth chamber & greenhouse	Provides resistance	[41]
CRISPR/ Cas9	Rice (*O. sativa*)	*SWEET11, SWEET13 & SWEET14*	Base deletion	Bacteria	Bacterial blight	Growth chamber & field	Provides broad-spectrum resistance	[42]
CRISPR/ Cas9	Tomato (*S. lycopersicum*)	*CCD8*	Base insertion/ deletion	Parasitic weed	Phelipanche aegyptiaca	Greenhouse & field	Develops host resistance	[43]
VIGS, RNAi & CRISPR/ Cas9	Wheat (*Triticum aestivum* L.)	*TaEDR1*	Base deletion	Fungus	Powdery mildew		Enhances resistance	[10]
CRISPR/ Cas9	Common tobacco (*Nicotiana tabacum*)		Base insertion/ deletion	Virus	Geminivirus		Defeat the mixed infections of geminivirus disease complexes	[44]
TALEN-based	Rice (*O. sativa*)	*OS11N3*	Base deletion	Bacteria	Bacterial blight	Growth chamber	Resistance to Bacterial Blight	[45]

GR, Gamma radiation; FNI, Fast-neutron irradiation; MNU, N-methyl-N-nitrosourea; VIGS, virus-induced gene silencing and CRISPR; EMS, Ethyl methane sulphonate; NaN3, sodium azide; UV, Ultraviolet mutagenesis.

Table 1.
Biotic-resistant mutant plants using gene editing, physical and chemical mutagens in two decades.

sulfate and diepoxybutane are effective chemical mutagens used [12]. Ionising radiations cause an array of chemical changes in crop plants by penetrating deep into plant tissues. Inversion, translocation, breakdown, and duplication of chromosomes, as well as point mutation, are among the chemical changes caused by ionising radiations [13]. Gamma rays are the most effective and widely used among ionising radiations in induced mutation [9]. Furthermore, chemical mutagens have also proven effective in inducing mutation. Their usage is easy, does not involve any specialised tools and generates a high rate of mutation. They effect changes in single base-pair or point mutations, or single-nucleotide polymorphisms (SNPs) [9]. Chemical mutagens are often used on vegetative propagules, seedlings, seeds, and tissues cultured in vitro such as explants of leaf and stem, ovules, anthers, microspores, and cell cultures, among others [14, 15]. Biological mutagens use living organisms such as bacteria (*Agrobacterium tumefaciens*, & *Agrobacterium rhizogenes*) and viruses (*Tobacco mosaic virus*) to cause mutation in DNA of organisms. The advancement of NGS strategies has been used to overcome random mutation. This is supported by genome editing tools namely, TALEN, CRISPR/Cas9, and ZFNs and they are applied in diverse plants. The effectiveness of these tools is well documented in different plants (**Table 1**). Currently, the combination of mutagenesis, PCR-based techniques and NGS provides the avenue to gene function contributing to crop improvement.

3. Increasing efficiency of targeted mutagenesis and identification of some novel traits

The evolution and advancement of targeted mutagenesis have significantly increased its efficiency by limiting the shortfall (error catastrophe and evolutionary escape) linked with the procedure. The quest for highly efficient targeted mutagenesis has resulted in small laboratories an access to targeted mutagenesis [46]. Targeted mutagenesis is achieved by embracing techniques such as Zinc Finger Nucleases (ZFNs), Transcription Activator Like Effector Nucleases (TALENs), and Clustered Regularly Interspaced Short Palindromic Repeats (CRISPR)/CRISPR-associated protein 9 (Cas9) [10, 47]. These techniques stimulate double-stranded breaks within the genomic DNA section of interest and consequently recombine [46, 48]. Setbacks within the application of ZFN include its domain assemble in binding with the nucleotides with high-affinity couple with target site selection [49]. However, off-target effects are reduced by using modified ZFNs and universal deep-learning models which now facilitated targeted mutagenesis [47, 50, 51]. Increasing the efficiency of TALENS requires heterodimer *FokI* domains and the use of an effective and efficient delivery system (electroporation or viral vectors) [52–54]. The clarity of design, versatility, and efficiency of CRISPR/Cas9 make it an acceptable and widely used gene editing technique [55, 56]. Thus, leading to advances in induced mutagenesis [57, 58]. Nonetheless, there are some challenges related to the low mutation efficiency in *Arabidopsis thaliana* and *Medicago truncatula* compared to the improvement attained in monocot species [59, 60]. Several strategies are designed in optimising CRISPR/Cas9. These include decreasing sgRNA-Cas9 concentration and its length; and truncated gRNAs which result in achieving high on-target efficiency [61–63]. Other strategies also include the development of Cas9 variant enzymes, efficient expression of several sgRNAs, and promotor-activated expression of Cas9 [63].

Plant species	Mutagen	Mutagenic significance	Reference
Hordeum vulgare L.	GR, X-ray & Ethylene imine	Enhances tolerance to low temperatures, drought, and frost	[67]
Brassica napus L. & *Brassica campestris* L.	GR	Tolerance to salinity	[67]
Lathyrus sativus L.	GR	Improves resistance to drought and high temperatures	[67]
Arachis hypogaea L. & *O. sativa*	GR	resistance to low temperature and drought	[67]
Triticum aestivum	GR & Beta rays	Increases resistance to low temperature, salinity, alkalinity and drought	[67]
Iris sp.	GR	Strengthen resistance to low temperatures	[67]
Phaseolus vulgaris L., *Vigna radiata* (L.) Wil. & *Cajanus cajan* Millsp.	GM	Tolerance to drought, salinity and phosphorus deficiency	[67]
O. sativa	GR	Tolerance to salinity, drought, and low pH	[67]
Setaria sp.	GR & FNI	Tolerance to drought	[67]
Brassica napus L.	GR & EMS	Tolerance to low temperatures and drought	[67, 68]
Glycine max	Laser & GR	Tolerance to drought	[67]
A. thaliana & *O. sativa*	EMS	Tolerance to salinity	[69, 70]
T. aestivum	NEU & EMS	Resistance to drought, low temperatures and salinity	[67, 71]
T. aestivum	EMS	Increases salt tolerance	[71]

GR, Gamma ray; EMS, Ethyl methane sulphonate; FNI, Fast-neutron irradiation; NEU, N-nitroso-N-ethyl urea.

Table 2.
Abiotic-resistant mutant varieties among different crops.

Mutation breeding presents breeders an opportunity to overcome plant deficiencies without changing their original identity [64]. Identification of novel traits is supported via induced mutagenesis that could otherwise not be possible in a natural gene pool [65]. This is enhanced by the advent of molecular techniques which increase crop improvement programmes [66]. Non-targeted mutagenesis approaches offer a high possibility of identifying novel traits. The adoption of mutagenesis has engendered the discovery of several unique traits within mutagenized populations of several crops (**Tables 1** and **2**).

4. Contributions of mutagenesis in plant improvement

Mutagenesis has contributed extensively to generating a pool of plant genetic resources (PGRs) resulting in the development of smart, climate-friendly varieties which contribute to enhancing farming resilience, and return on investment for farmers. Mutant varieties have proven to have desirable traits such as tolerance to stresses (biotic and abiotic), higher yields, biofortification and generally, improvement of the genetic composition of plants (**Tables 1** and **2**).

4.1 Plant biodiversity and genetic resources

There is a constant loss of agricultural PGRs with approximately 37% of them going extinct [72]. Thus, the genetic base of most crops is narrowing and makes them succumb to environmental stresses. Plant biodiversity is key in limiting these consequences and induced mutation plays a vital role in supporting breeder to generate diverse genetic resources to mitigate current climate variations to achieve food security [73–76]. Mutagenesis offers breeders a tool for generating crop variability in the shortest time as compared to traditional crosses [77]. Consequently, establishing and maintaining PGRs is pivotal to crop improvement. Mutation breeding allows plant breeders to work with farmers to breed varieties of crops that are high-yielding and more resistant to disease, resulting in the intensification of crop production (**Tables 1** and **2**). Breeding programs utilised PGRs generated via mutagenesis by increasing the gene pools for breeding purposes, development of genetic stocks with desirable traits, characterisation of PGRs, and cultivar development by improving farmers' preferred cultivars with desirable traits. The availability of PGRs would contribute to overcoming the various challenges affecting plant growth and development. For instance, mutagenesis has resulted in creating 2577 mutant varieties only in 10 top countries across the globe the highest number of crop mutants developed (**Figure 1**) [67]. China and Pakistan have contributed to producing 817 and 59 mutant varieties respectively [67]. Similarly, several crops have received much attention in terms of induced mutagenesis. For example, 873 rice mutants have been developed, followed by barley (307 mutant), Chrysanthemum (285 mutants), wheat (265 mutants), soybean (182 mutant) and the 10th rank Dahlia (36 mutant crops) globally (**Figure 2**) [67].

4.2 Biotic stress resistance

Population increase coupled with modernization usually leads to significant changes in the environmental forces that influence the production of agricultural goods. The environmental forces mainly arise from biotic and abiotic elements.

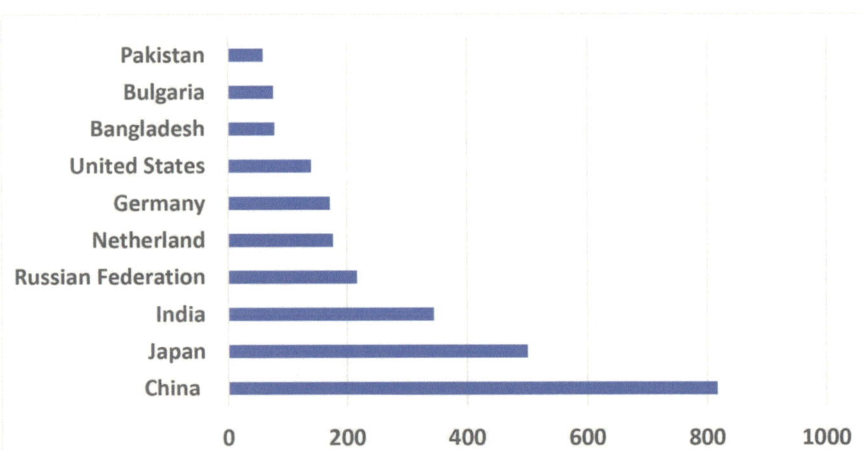

Figure 1.
Top 10 countries in the world highest number of mutants crops developed [67].

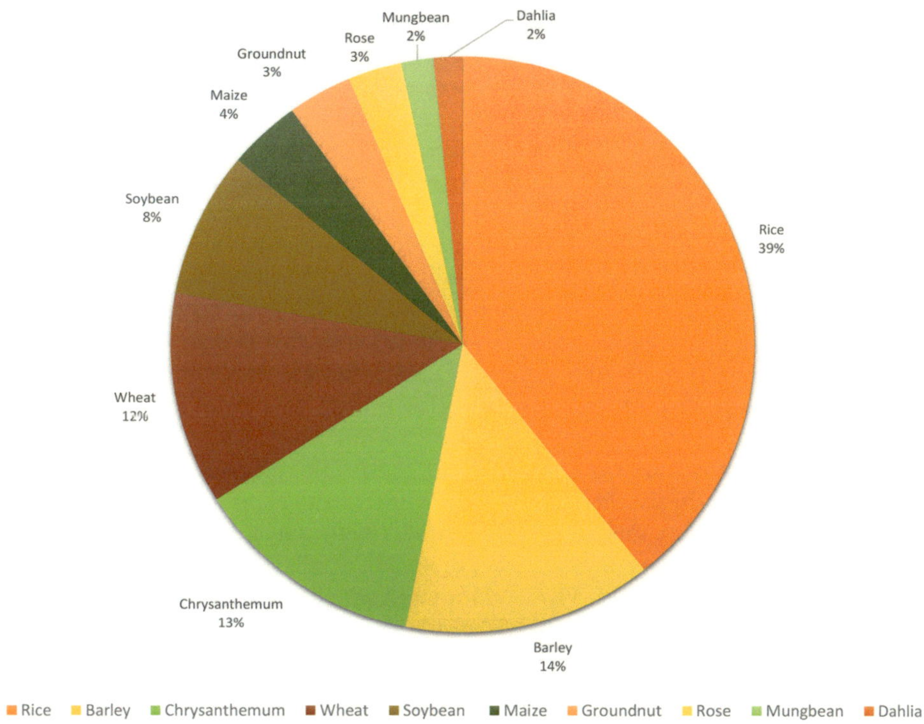

Figure 2.
Top 10 crops having highest number of crop mutants [67].

The biotic force arises from the actions of fungi, bacteria, viruses, nematodes, and pests, among others. These trigger stress by impeding plants' growth and development, eventually leading to reduction in yield as well as value. Therefore, it is imperative for crop improvement strategies to be used to devise approaches to counter the effects of biotic stress. Hence, the application of mutation breeding in overcoming biotic stress is crucial and it has proven to be successful. Plant breeders improve plants by generating plant genetic resources by mimicking the process of spontaneous mutation that exists in nature. Exposure of plants to stress triggers a complex and redundancy signals, perception and expression levels crosstalk among several pathways in order to overcome the stress. This demands an array of plants with varied genetic backgrounds, biochemical and metabolic variations to generate alleles required for engineering plants to tolerate biotic stresses.

Stress caused by disease-causing organisms in plants has witnessed improvement in combating them. A number of plants are developed to offer resistance to particular pathogens and insects. The durability of this tolerance is questioned due to the development of biotypes of pathogen strains stimulating breeders to generate new resistance against the biotic stresses that have newly emerged [78]. Many biotic stresses such as fungi, viruses, bacteria, nematodes, and insects interrupt the structure of plants. They are known for causing about 20–40% losses in agricultural production [79]. Specifically, in leguminous crops such as chickpea, common bean, faba bean and alfalfa, *Sclerotinia sclerotiorum* causes yield loss of up to 100% whiles affecting their seed quality and the entire plant function [80]. To mitigate the problem of hunger, techniques that increase pathogens' resistance to host crops to reduce their damage to crop production are required [81]. Plant scientists have developed crops against most of these stresses via mutagenesis approaches (**Table 1**). This is facilitated by plant molecular genetics making induced mutation a highly efficient technique for studies in plants leading to introgression of desirable traits in several crops such as rice, soybean, wheat, and sesame [24]. Although classical plant breeding has engendered the development of novel traits in plants, however, it is time-consuming. On the contrary, mutation breeding is affordable, safe and has no legal restriction on its application. Induced mutagenesis is facilitated by the evolving next-generation sequencing (NGS) and its non-destructive evaluation systems limiting the time frame, and labour needed for mutation breeding. It is undoubted that, for decades to come, induced mutation will retain its value in plant science especially when complex trait improvements such as diseases, and pests are involved. Currently, mutation techniques ranging from physical, chemical, gene editing (CRISPR/Cas9, TALEN-based, VIGS, RNAi interference) are employed in combating biotic stresses conforming plants. For instance, 10 different disease stresses in rice (*Oryza sativa* L.) are improved via mutagenesis using different approaches (**Table 1**). Application of EMS in soybean (*Glycine max* L.) has resorted in the development of Chlorsulfuron-Resistant soybean [27]. Resistance of grapevine (*Vitis vinifera*), and tomato (*Solanum lycopersicum* L.) to powdery mildew diseases caused by fungus have been reported to be promoted via induced mutagenesis (**Table 1**) [36, 38]. Similarly, bacteria blight disease in rice (*O. sativa*) is reduced by CRISPR/Cas9 mutagenesis through base insertion/deletion of genes (*OsSWEET11, 13* and *14*; *OsXa13/Os8N3*; *OsAvrXa7*) (**Table 1**) [35, 39, 40]. Data shows the power of gene editing and physical mutagens in recent decades in improving plants' tolerance/resistance to biotic stresses (**Table 1**).

4.3 Abiotic stress resistance

The sessile nature of plants makes them very susceptible to the adverse effects of abiotic stress such as cold, heat, flood, drought, salinity, heavy metals, reduced or excessive UV radiation, acidity, alkalinity, and nutrient-deficient soils [82, 83]. Abiotic stress happens to be the primary cause of crop loss globally and leads to an annual yield loss of over 50% in major crops [84].

In our quest to sustain agricultural production and food supply, numerous attempts are being made to generate mutant varieties that can withstand abiotic stress in light of climate change, with the use of various plant-breeding techniques. Currently, mutation breeding has successfully generated superior varieties of legumes, grains and cereals, roots and tubers, cotton, and sugarcane that are high yielding, and tolerant to some abiotic stresses. The mutant variety database (MVD) hosts 3402 mutants of which around 160 mutant varieties are tolerant to abiotic stresses [67]. For instance, 'Zhefu 802', a mutant rice variety developed in China is resistant to low temperatures [85]. In Pakistan, induced mutation in Basmati 370 rice generated a new variety "Kashmir Basmati" that has tolerance to cold, matures early and retains parental traits such as cooking quality and aroma [86]. **Table 2** presents a highlight of breakthroughs made through induced mutation to improve resistance to abiotic stresses of different crops.

4.4 Nutritional and quality trait improvement

The nutritional qualities of crops are of concerns to humankind. In the worldwide efforts to feed a growing and nutritional demanding human population, mutation breeding is a crucial strategy [87]. Several mutational breeding has targeted enhancing phytonutrients, essential minerals, proteins, and oil for humans and animals. Mutation breeding have played a crucial component in developing variant cultivars having spurious modifications in genes using insertion mutagenesis such as T-DNA insertion in rice, transposon or retrotransposon tagging in maize or rice, and chemical/irradiation mutagenesis that produces new characters for crop improvement [88]. Classical examples are the mutant varieties namely Zornitsa, Madan, and NIFA-Mustard Canola developed with enhanced oil and protein content [67]. Similarly, the nutritional component of rice mutants has been improved compared to its wild type [89]. The development of mutant crops from soybean, rice, wheat and barley have witnessed increases in their bioavailability of nutrients and minerals [90]. The ratio of oleic to linoleic acid has been enhanced and the palmitic acid content in peanut has been reduced through induced mutation [91]. The β-carotene levels in pepper fruits were increased by creating mutants through the application of physical and chemical mutagens [92]. Similarly, in rapeseed application of EMS result in improve its genetic composition in terms of oil content and fatty acids [93].

5. Future of mutagenesis for improving plants

Induced mutation has proven essential for close to 100 years generating genetic variations and subsequent release of mutant cultivars. Bottlenecks to mutation breeding arise from its lethal, low rate, laborious screening process, and most mutations being recessive, among others. Radiation mutation breeding now takes advantage of space projects as this seems to become common in the future. Specifically, the space environment denotes outer space outside the atmosphere and is accompanied by microgravity, radiation, and alternating magnetic fields. This field will offer more openings for mutation breeding as space breeding provides a high rate of mutation, in several directions [94]. Aerospace (space) mutation is applied in developing a new variety [95]. Crop space breeding studies have been carried out in China since 1987 and shreds of evidence showed have been encouraging resulting in the release of 66 mutant varieties, which include crops such as tomato, pepper, rice, wheat and sesame [96]. This is evident that, in the future countries can tap into the space breeding procedures. This is due to its associated benefits such as resulting inducing genetic changes in the seed of crops and chromosomal aberrations surge in the seeds attributed to the combined effects of cosmic radiation and microgravity. More efforts will concentrate on cooperating mutagenesis, PCR-based methods, and mapping techniques (NGS techniques) to explore further the function of key genes and to embrace integrative platforms to advance functional genomics innovations.

6. Conclusion

Induced mutagenesis is key to contributing to food security by developing climate-smart crops. Hence, it represents a tool in fighting against hunger, and malnutrition whiles increasing farmers' profit margins. The usage of mutagens such as chemical agents is easy and does not require high skills making it user-friendly. The key challenges in mutation breeding are the low efficiency in targeted mutagenesis resulting in off-target and random mutation. Current efforts are geared towards embracing PCR-based methods, and NGS techniques in induced mutagenesis. This integration offers the opportunity for reverse breeding and to explore keys and their roles in plants. In summary, multi-omics and precise mutagenesis interventions improve plant yields and qualities by developing climate-smart crops.

Conflict of interest

The authors declare no conflict of interest.

Author details

Augustine Antwi-Boasiako[1,2]*, Padmore Adu-Antwi[1], Richard Adu Amoah[3], Augustine Boakye Boateng[1], Joseph Gyau[1], Matilda Frimpong[1], Isaac Newton Boakye-Mensah[1] and Ivy Odi Ahiamadia[4]

1 Crops Research Institute, Council for Scientific and Industrial Research, Kumasi, Ghana

2 National Center for Soybean Improvement, Key Laboratory of Biology and Genetics and Breeding for Soybean, Ministry of Agriculture, State Key Laboratory of Crop Genetics and Germplasm Enhancement, Nanjing Agricultural University, Nanjing, China

3 Plant Genetic Resources Research Institute, Council for Scientific and Industrial Research, Bunso, Ghana

4 Cocoa Health and Extension Division, Ghana Cocoa Board, Accra, Ghana

*Address all correspondence to: bbbantwi@cropsresearch.org

References

[1] Sattar MN, Iqbal Z, Al-Khayri JM, Jain SM. Induced genetic variations in fruit trees using new breeding tools: Food security and climate resilience. Plants. 2021;**10**(7):1347. DOI: 10.3390/plants10071347

[2] Mba C, Afza R, Bado S, Jain SM. Induced mutagenesis in plants using physical and chemical agents. Plant Cell Culture: Essential Methods. 2010;**20**:111-130

[3] Kashtwari M, Wani AA, Dhar MK, Jan S, Kamili AN. Development of an efficient in vitro mutagenesis protocol for genetic improvement of saffron (Crocus sativus L.). Physiology and Molecular Biology of Plants. 2018;**24**:951-962. DOI: 10.1007/s12298-018-0576-6

[4] Wijerathna-Yapa A, Ramtekey V, Ranawaka B, Basnet BR. Applications of in vitro tissue culture technologies in breeding and genetic improvement of wheat. Plants. 2022, 2022;**11**(17):2273. DOI: 10.3390/plants11172273

[5] Sherpa R, Devadas R, Bolbhat SN, Nikam TD, Penna S. Gamma radiation induced In-vitro mutagenesis and isolation of mutants for early flowering and Phytomorphological variations in dendrobium 'Emma white'. Plants. 2022;**11**(22):3168. DOI: 10.3390/plants11223168

[6] Penna S, Vitthal SB, Yadav PV. In vitro mutagenesis and selection in plant tissue cultures and their prospects for crop improvement. Bioremediation, Biodiversity, Bioavailability. 2012;**6**:6-14

[7] Oladosu Y, Rafii MY, Abdullah N, Hussin G, Ramli A, Rahim HA. Principle and application of plant mutagenesis in crop improvement: A review. Biotechnology & Biotechnological Equipment. 2016;**30**(1):1-16. DOI: 10.1080/13102818.2015.1087333

[8] Joung JK, Sander JD. TALENs: A widely applicable technology for targeted genome editing. Nature reviews Molecular Cell Biology. 2013;**14**(1):49-55. DOI: 10.1038/nrm3486

[9] Suprasanna P, Mirajkar SJ, Patade VY, Jain SM. Induced mutagenesis for improving plant abiotic stress tolerance. In: Mutagenesis: Exploring Genetic Diversity of Crops. The Netherlands: Wageningen Acdemic Publishers; 2014. pp. 345-376. DOI: 10.3920/978-90-8686-796-7_17

[10] Zhang Y, Bai Y, Wu G, Zou S, Chen Y, Gao C, et al. Simultaneous modification of three homoeologs of Ta EDR 1 by genome editing enhances powdery mildew resistance in wheat. The Plant Journal. 2017;**91**(4):714-724. DOI: 10.1111/tpj.13599

[11] Brauer EK, Balcerzak M, Rocheleau H, Leung W, Schernthaner J, Subramaniam R, et al. Genome editing of a deoxynivalenol-induced transcription factor confers resistance to Fusarium graminearum in wheat. Molecular Plant-Microbe Interactions. 2020;**33**(3):553-560

[12] Sikora P, Chawade A, Larsson M, Olsson J, Olsson O. Mutagenesis as a tool in plant genetics, functional genomics, and breeding. International Journal of Plant Genomics. 2011:1-13. DOI: 10.1155/2011/314829

[13] Britt AB. DNA damage and repair in plants. Annual Review of Plant Physiology and Plant Molecular Biology. 1996;**45**:75-100. DOI: 10.1146/annurev.arplant.47.1.75

[14] Suprasanna P, Sonawane BV, Patade VY. In vitro mutagenesis and selection: Strategies and prospects for crop improvement. In: Kozgar MI, Khan S, editors. Induced Mutagenesis in Crop Plants. Bioremediation, Biodiversity and Bioavailability. Vol. 6 (Special Issue 1). Tokyo, Japan: Global Science Books; 2012. pp. 6-14

[15] Xu L, Najeeb U, Naeem MS, Wan GL, Jin ZL, Khan F, et al. Technological innovations in major world oil crops. In: Gupta SK, editor. In Vitro Mutagenesis and Genetic Improvement. Vol. 2. New York, NY, USA: Springer; 2012. pp. 151-173

[16] Abdel-Hamid AA, Galal OA, Elmoghazy A, Abd Elrazek MM. Using gamma rays for genetic improvement of Rice resistance to blast disease. Journal of Agricultural Chemistry and Biotechnology. 2022;**13**(10):85-89. DOI: 10.21608/jacb.2022.152986.1028

[17] Gajbar TD, Kamble M, Adhikari S, Konappa N, Satapute P, Jogaiah S. Gamma-irradiated fenugreek extracts mediates resistance to rice blast disease through modulating histochemical and biochemical changes. Analytical Biochemistry. 2021;**618**:14121. DOI: 10.1016/j.ab.2021.114121

[18] Kara A, Arici ŞE. Determination of gamma rays efficiency against Rhizoctonia solani in potatoes. Open Chemistry. 2019;**17**(1):254-259. DOI: 10.1515/chem-2019-0033

[19] Kumari V, Singh A, Chaudhary HK, Kumar PR, Jambhulkar S, Sanju S. Identification of Phytophthora blight resistant mutants through induced mutagenesis in sesame (Sesamum indicum L.). Indian Phytopathology. 2019;**72**:71-77. DOI: 10.1007/s42360-018-0089-9

[20] Bart RS, Chern M, Vega-Sánchez ME, Canlas P, Ronald PC. Rice Snl6, a cinnamoyl-CoA reductase-like gene family member, is required for NH1-mediated immunity to Xanthomonas oryzae pv. Oryzae. PLoS Genetics. 2010;**6**(9):e1001123. DOI: 10.1371/journal.pgen.1001123

[21] Shah TM, Atta BM, Mirza JI, Haq MA. Screening of chickpea (Cicer arietinum) induced mutants against Fusarium wilt. Pakistan Journal of Botany. 2009;**41**(4):1945-1955

[22] Peiris R, Wickramasinghe TK, Indrasena SP. M 127-a promising tomato variety developed through induced mutation technique. In: Induced Plant Mutations in the Genomics Era. Proceedings of an International Joint FAO/IAEA Symposium. Rome: Food and Agriculture Organization of the United Nations; 2008. pp. 379-380

[23] Zhang MX, Xu JL, Luo RT, Shi D, Li ZK. Genetic analysis and breeding use of blast resistance in a japonica rice mutant R917. Euphytica. 2003;**130**:71-76. DOI: 10.1023/A:1022380626371

[24] Zambrano AY, Demey JR, Fuchs M, González V, Rea R, De Sousa O, et al. Selection of sugarcane plants resistant to SCMV. Plant Science. 2003;**165**(1):221-225. DOI: DOI.org/10.1016/S0168-9452(03)00162-6

[25] Kunz BA, Dando PK, Grice DM, Mohr PG, Schenk PM, Cahill DM. UV-induced DNA damage promotes resistance to the biotrophic pathogen Hyaloperonospora parasitica in Arabidopsis. Plant Physiology. 2008;**148**(2):1021-1031. DOI: 10.1104/pp.108.125435

[26] Gunasekera TS, Paul ND, Ayres PG. The effects of ultraviolet-B (UV-B: 290-320 nm) radiation on blister blight disease of tea (Camellia sinensis). Plant Pathology. 1997;**46**(2):179-185. DOI: 10.1046/j.1365-3059.1997.d01-216.x

[27] Ustun R, Uzun B. Development of a high yielded Chlorsulfuron-resistant soybean (Glycine max L.) variety through mutation breeding. Agriculture. 2023;**13**(3):559. DOI: 10.3390/agriculture13030559

[28] Potupureddi G, Balija V, Ballichatla S, Gokulan CG, Awalellu K, Lekkala S, et al. Mutation resource of samba Mahsuri revealed the presence of high extent of variations among key traits for rice improvement. PLoS One. 2021;**16**(10):e0258816. DOI: 10.1371/journal.pone.0258816

[29] Lo KL, Chen YN, Chiang MY, Chen MC, Panibe JP, Chiu CC, et al. Two genomic regions of a sodium azide induced rice mutant confer broad-spectrum and durable resistance to blast disease. Rice. 2022;**15**:1-16

[30] Ge H, Li Y, Fu H, Long G, Luo L, Li R, et al. Production of sweet orange somaclones tolerant to citrus canker disease by in vitro mutagenesis with EMS. Plant Cell, Tissue and Organ Culture. 2015;**123**:29-38

[31] Kiruki S, Onek LA, Limo M. Azide-based mutagenesis suppresses Striga hermonthica seed germination and parasitism on maize varieties. African Journal of Biotechnology. 2006;**5**(10):866-870

[32] Huang Q, Lin B, Cao Y, Zhang Y, Song H, Huan C, et al. CRISPR/Cas9-mediated mutagenesis of the susceptibility gene OsHPP04 in rice confers enhanced resistance to rice root-knot nematode. Frontiers. Plant Science. 2023;**14**. DOI: 10.3389/fpls.2023.1134653

[33] Fan S, Zhang Z, Song Y, Zhang J, Wang P. CRISPR/Cas9-mediated targeted mutagenesis of GmTCP19L increasing susceptibility to Phytophthora sojae in soybean. PLoS One. 2022;**17**(6):e0267502. DOI: 10.1371/journal.pone.0267502

[34] Thomazella DP, Seong K, Mackelprang R, Dahlbeck D, Geng Y, Gill US, et al. Loss of function of a DMR6 ortholog in tomato confers broad-spectrum disease resistance. Proceedings of the National Academy of Sciences. 2021;**118**(27):e2026152118

[35] Zafar K, Khan MZ, Amin I, Mukhtar Z, Yasmin S, Arif M, et al. Precise CRISPR-Cas9 mediated genome editing in super basmati rice for resistance against bacterial blight by targeting the major susceptibility gene. Frontiers in Plant Science. 2020;**11**:575

[36] Martínez MIS, Bracuto V, Koseoglou E, Appiano M, Jacobsen E, Visser RG, et al. CRISPR/Cas9-targeted mutagenesis of the tomato susceptibility gene PMR4 for resistance against powdery mildew. BMC Plant Biology. 2020;**20**(1):1-13. DOI: 10.1186/s12870-020-02497-y

[37] Yoon YJ, Venkatesh J, Lee JH, Kim J, Lee HE, Kim DS, et al. Genome editing of eIF4E1 in tomato confers resistance to pepper mottle virus. Frontiers in Plant Science. 2020;**11**:1098

[38] Wan DY, Guo Y, Cheng Y, Hu Y, Xiao S, Wang Y, et al. CRISPR/Cas9-mediated mutagenesis of VvMLO3 results in enhanced resistance to powdery mildew in grapevine (Vitis vinifera). Horticulture Research. 2020;**7**:116. DOI: 10.1038/s41438-020-0339-8

[39] Xu Z, Xu X, Gong Q, Li Z, Li Y, Wang S. Engineering broad-spectrum bacterial blight resistance by simultaneously disrupting variable TALE-binding elements of multiple susceptibility genes in rice. Molecular Plant. 2019;**12**(11):1434-1446. DOI: DOI.org/10.1016/j.molp.2019.08.006

[40] Kim YA, Moon H, Park CJ. CRISPR/Cas9-targeted mutagenesis of Os8N3 in rice to confer resistance to Xanthomonas oryzae pv. Oryzae. Rice. 2019;**12**(1):1-3

[41] Ortigosa A, Gimenez-Ibanez S, Leonhardt N, Solano R. Design of a bacterial speck resistant tomato by CRISPR/Cas9-mediated editing of Sl JAZ 2. Plant Biotechnology Journal. 2019;**17**(3):665-673

[42] Oliva R, Ji C, Atienza-Grande G, Huguet-Tapia JC, Perez-Quintero A, Li T. Broad-spectrum resistance to bacterial blight in rice using genome editing. Nature Biotechnology. 2019;**37**(11):1344-1350. DOI: 10.1038/s41587-019-0267-z

[43] Bari VK, Nassar JA, Kheredin SM, Gal-On A, Ron M, Britt A, et al. CRISPR/Cas9-mediated mutagenesis of CAROTENOID CLEAVAGE DIOXYGENASE 8 in tomato provides resistance against the parasitic weed Phelipanche aegyptiaca. Scientific Reports. 2019;**9**(1):11438. DOI: 10.1038/s41598-019-47893-z

[44] Baltes NJ, Hummel AW, Konecna E, Cegan R, Bruns AN, Bisaro DM, et al. Conferring resistance to geminiviruses with the CRISPR–Cas prokaryotic immune system. Nature Plants. 2015;**1**(10):1-4

[45] Li T, Liu B, Spalding MH, Weeks DP, Yang B. High-efficiency TALEN-based gene editing produces disease-resistant rice. Nature Biotechnology. 2012;**30**(5):390-392. DOI: 10.1038/nbt.2199

[46] Hunter CT. CRISPR/Cas9 targeted mutagenesis for functional genetics in maize. Plants. 2021;**10**(4):723. DOI: 10.3390/plants10040723

[47] Gaj T, Guo J, Kato Y, Sirk SJ, Barbas CF. Targeted gene knockout by direct delivery of zinc-finger nuclease proteins. Nature Methods. 2012;**9**(8):805-807. DOI: 10.1038/nmeth.2030

[48] Mosbach V, Poggi L, Viterbo D, Charpentier M, Richard GF. TALEN-induced double-strand break repair of CTG trinucleotide repeats. Cell Reports. 2018;**22**(8):2146-2159. DOI: 10.1016/j.celrep.2018.01.083

[49] Ramirez CL, Certo MT, Mussolino C, Goodwin MJ, Cradick TJ, McCaffrey AP, et al. Engineered zinc finger nickases induce homology-directed repair with reduced mutagenic effects. Nucleic Acids Research. 2012;**40**(12):5560-5568. DOI: 10.1093/nar/gks179

[50] Doyon Y, Vo TD, Mendel MC, Greenberg SG, Wang J, Xia DF, et al. Enhancing zinc-finger-nuclease activity with improved obligate heterodimeric architectures. Nature Methods. 2011;**8**(1):74-79

[51] Ichikawa DM, Abdin O, Alerasool N, Kogenaru M, Mueller AL, Wen H, et al. A universal deep-learning model for zinc finger design enables transcription factor reprogramming. Nature Biotechnology. 2023:1-13. DOI: 10.1038/s41587-022-01624-4

[52] Holkers M, Maggio I, Liu J, Janssen JM, Miselli F, Mussolino C, et al. Differential integrity of TALE nuclease genes following adenoviral and lentiviral vector gene transfer into human cells. Nucleic Acids Research. 2013;**41**(5):e63-e63. DOI: 10.1093/nar/gks1446

[53] Suzuki K, Yu C, Qu J, Li M, Yao X, Yuan T, et al. Targeted gene correction minimally impacts whole-genome mutational load in human-disease-specific induced pluripotent stem cell clones. Cell Stem Cell. 2014;**15**(1):31-36. DOI: 10.1016/j.stem.2014.06.016

[54] Yang L, Guell M, Byrne S, Yang JL, De Los AA, Mali P, et al. Optimization of scarless human stem cell genome editing. Nucleic Acids Research. 2013;**41**(19):9049-9061. DOI: 10.1093/nar/gkt555

[55] Wu X, Kriz AJ. Sharp PA. Target specificity of the CRISPRCas9 system. Quantitative Biology. 2014;2(2):59-70. DOI:10.1007/s40484-014-0030-x

[56] Bortesi L, Fischer R. The CRISPR/Cas9 system for plant genome editing and beyond. Biotechnology Advances. 2015;**33**(1):41-52. DOI: 10.1016/j.biotechadv.2014.12.006

[57] Rao YC, Yang X, Pan CY, Wang C, Wang KJ. Advance of clustered regularly interspaced short palindromic repeats-Cas9 system and its application in crop improvement. Frontiers in Plant Science. 2022;**13**:839001. DOI: 10.3389/fpls.2022.839001

[58] Liu T, Zhang X, Li K, Yao Q, Zhong D, Deng Q, et al. Large-scale genome editing in plants: Approaches, applications, and future perspectives. Current Opinion in Biotechnology. 2023;**79**:102875. DOI: 10.1016/j.copbio.2022.10287

[59] Feng Z, Zhang Z, Hua K, Gao X, Mao Y, Botella JR, et al. A highly efficient cell division-specific CRISPR/Cas9 system generates homozygous mutants for multiple genes in Arabidopsis. International Journal of Molecular Sciences. 2018;**19**(12):3925. DOI: 10.3390/ijms19123925

[60] Castel B, Tomlinson L, Federica L, Yang Y, Jones JDG. Optimization of T-DNA architecture for Cas9-mediated mutagenesis in Arabidopsis. PLoS One. 2019;**14**(1):e0204778. DOI: 10.1371/journal.pone.0204778

[61] Fu Y, Sander JD, Reyon D, Cascio VM, Joung JK. Improving CRISPR–Cas nuclease specificity using truncated guide RNAs. Nature Biotechnology. 2014;**32**:279-284. DOI: 10.1038/nbt.2808

[62] Liu XJ, Yang JT, Song YY, Zhang XC, Wang XJ, Wang ZX. Effects of sgRNA length and number on gene editing efficiency and predicted mutations generated in rice. Crop Journal. 2022;**10**:577-581. DOI: 10.1016/j.cj.2021.05.015

[63] Zhou J, Luan X, Liu Y, Wang L, Wang J, Yang S. Strategies and methods for improving the efficiency of CRISPR/Cas9 gene editing in plant molecular breeding. Plants. 2023;**12**(7):1478. DOI: 10.3390/plants12071478

[64] Sao R, Sahu PK, Patel RS, Das BK, Jankuloski L, Sharma D. Genetic improvement in plant architecture, maturity duration and agronomic traits of three traditional Rice landraces through gamma ray-based induced mutagenesis. Plants. 2022;**11**(24):3448. DOI: 10.3390/plants11243448

[65] Roychowdhury R, Tah J. Mutagenesis—A potential approach for crop improvement. In: Hakeem K, Ahmad P, Ozturk M, editors. Crop Improvement. Boston, MA: Springer; 2013. DOI: 10.1007/978-1-4614-7028-1_4

[66] Jain SM. In vitro mutagenesis in banana (Musa spp.) improvement. Acta Horticulturae. 2010;**879**:605-614. DOI: 10.17660/ActaHortic.2010.879.67

[67] FAO/IAEA. Mutant Variety Database. Available from: https://mvd.iaea.org/ [Accessed: 30th March 2023]

[68] Channaoui S, Mazouz H, Labhilili M, El Fechtali M, Nabloussi A. Inheritance of dwarfism and narrow lobed-leaf in two rapeseed (Brassica napus L.) mutant lines. Heliyon. 2022;**8**(12):e12649. DOI: 10.1016/j.heliyon.2022.e12649

[69] Takagi H, Tamiru M, Abe A, Yoshida K, Uemura A, Yaegashi H, et al. MutMap accelerates breeding of a salt-tolerant rice cultivar. Nature Biotechnology. 2015;**33**(5):445-449

[70] Shankar A, Choudhary OP, Bhatia D. Effect of EMS induced mutation in rice cultivar Nagina 22 on salinity tolerance. bioRxiv. 2021:2021-08

[71] Lethin J, Shakil SSM, Hassan S, Sirijovski N, Töpel M. Olsson O, et al. Development and characterization of an EMS-mutagenized wheat population and identification of salt-tolerant wheat lines. BMC Plant Biology. 2020;**20**(1):1-15. DOI: 10.1186/s12870-019-2137-8

[72] Thomas DC, Alison C, Green E, Bakkenes M, Beaumont J, Collingham C, et al. Extinction risk from climate change. Nature. 2004;**427**(6970):145-148

[73] Adamu A, Aliyu H. Morphogical effects of sodium azide on tomato (Lycopersicon esculentum mill). Scientific World Journal. 2007;**2**(4). DOI: 10.4314/swj.v2i4.51755

[74] Kozgar MI, Khan S, Wani MR. Variability and correlations studies for total iron and manganese contents of chickpea (Cicer arietinum L.) high yielding mutants. American Journal of Food Technology. 2012;**7**:437-444. DOI: 10.3923/ajft.2012.437.444

[75] Ogwu MC, Osawaru ME, Ahana CM. Challenges in conserving and utilizing plant genetic resources (PGR). International Journal of Genetics and Molecular Biology. 2014;**6**:16-22. DOI: 10.5897/IJGMB2013.0083

[76] Salgotra RK, Chauhan BS. Genetic diversity, conservation, and utilization of plant genetic resources. Genes. 2023;**14**(1):174. DOI: 10.3390/genes14010174

[77] Ansari BS, Raina AA, Amin R, Jahan R, Malik S, Khan S. Mutation breeding for quality improvement: A case study for oilseed crops. In: Stephenson editor. Mutagensis Cytotoxicity. UK: Newcastle; 2021. pp. 171-221

[78] Hussain B. Modernization in plant breeding approaches for improving biotic stress resistance in crop plants. Turkish Journal of Agriculture and Forestry. 2015;**39**:515-530. DOI: 10.3906/tar-1406-176

[79] Walker PT. Quantification and economic assessment of crop losses due to pests, diseases and weeds. In: Advancing Agricultural Production in Africa: Cabs First Scientific Conference. Arusha, Tanzania: Elsevier; 1984

[80] Antwi-Boasiako A, Wang Y, Dapaah HK, Zhao T. Mitigating against Sclerotinia diseases in legume crops: A comprehensive review. Agronomy. 2022;**12**(12):3140. DOI: 10.3390/agronomy12123140

[81] Borrelli VMG, Brambilla V, Rogowsky P, Marocco A, Lanubile A. The enhancement of plant disease resistance using CRISPR/ Cas9 technology. Frontiers in Plant Science. 2018;**9**:1245. DOI: 10.3389/fpls.2018.01245

[82] He M, He CQ, Ding NZ. Abiotic stresses: General defenses of land plants and chances for engineering multistress tolerance. Frontiers in Plant Science. 2018;**9**:1771. DOI: 10.3389/fpls.2018.01771

[83] Mehta K. Effect of abiotic and biotic stress on the plant. Botanical Sciences. 2018;**1**:05-09

[84] Wang W, Vinocur B, Altman A. Plant responses to drought, salinity and extreme temperatures: Towards genetic engineering for stress tolerance. Planta. 2003;**218**:1-4

[85] Kharkwal MC, Shu QY. The role of induced mutations in world food security. In: Shu QY, editor. Induced Plant Mutations in the Genomics Era. Rome: Food and Agriculture Organization of the United Nations, Italy; 2009. pp. 33-38

[86] Awan MA. Use of induced mutations for crop improvement in Pakistan. Plant Mutation Breeding for Crop Improvement. 1991;**1**:67-72

[87] Raina A, Laskar RA, Khursheed S, Amin R, Tantray YR, Parveen K, et al. Role of mutation breeding in crop improvement-past, present and future. Asian Research Journal of Agriculture. 2016;**2**(2):1-3

[88] Farré G, Blancquaert D, Capell T, Van Der Straeten D, Christou P, Zhu CJAROPB. Engineering Complex Metabolic Pathways in Plants. 2014;**65**:187-223

[89] Zhang L, Shu L, Wang Y, Lu J, Shu Y, Wu X. Characterization of indicia-type giant embryo mutant rice enriched with nutritional components. Cereal Research Communications. 2007;**35**:1459-1468

[90] Jain MS, Suprasanna P. Induced mutations for enhancing nutrition and food production. Gene Conserve. 2011;**40**:201-215

[91] Kamdar JH, Jasani MD, Ajay BC, Rani K, Manivannan N, Vasanthi RP, et al. Fatty acid desaturase-2 (ahFAD2) mutant alleles in peanut (Arachis hypogaea L.) pre-breeding lines: An insight into the source, features, discourse, and selection of novel pre-breeding lines. Genetic Resources and Crop Evolution. 2021;**68**:529-549. DOI: 10.1007/s10722-020-00999-0

[92] Tomlekova N, Todorova V, Petkova V, Yancheva S, Nikolova V, Panchev I, et al. Creation and evaluation of induced mutants and valuable tools for pepper breeding programmes. In: Induced Plant Mutations in the Genomics Era. Vol. 1. Rome, Italy: Food and Agriculture Organization of the United Nations; 2009. pp. 187-190

[93] Van Zelm E, Zhang Y, Testerink C. Salt tolerance mechanisms of plants. Annual Review of Plant Biology. 2020;**71**:403-433. DOI: 10.1146/annurev-arplant-050718- 100005

[94] Liang X, Huang X, Mo M. Feasibility analysis of mulberry breeding by aerospace mutation. Guangxi Sericulture. 2009;**46**:32-36

[95] Chen Z, Wang H. Peiza Taifeng and Huahang No. 1. Crop Research. 2004;**4**:283. DOI: 10.16848/j.cnki.issn.1001-5280.2004.04.086

[96] Liu LX, Guo HJ, Zhao LS, Wang J, Gu JY, Zhao SR. Achievements and perspectives of crop space breeding in China. Induced Plant Mutations in the Genomics Era. 2009;**42**(1):213-215

Chapter 4

Genome Editing in Potato Using CRISPR/Cas Technology: Applications and Challenges

Deepa Beniwal, Shivani Chauhan and Harnoor Kaur Dhillon

Abstract

After rice and wheat, potato is the third most important food crop for human consumption. In Europe and several parts of America, potato is consumed predominantly. Like other vegetable crops, potato is prone to several biotic and abiotic stresses and due to climate change, such stresses are getting worse and affecting the yield and quality of harvested product. Both conventional breeding and transgenic approaches are being utilized to enhance the crop production by protecting the crop for different biotic and abiotic stresses. Genome editing technologies such as ZFNs and TALENs were earlier utilized for crop improvement. But recently, RNA-guided nuclease called CRISPR technology is in use for crop improvement. In potato, CRISPR/Cas is utilized for phenotyping, tuber quality, late blight resistance, potato virus Y resistance, herbicide tolerance, starch quality and biosynthesis, enzymatic browning, phosphate transport to roots and several other desirable traits. In this chapter, we summarize the information about major genome editing approaches and use of CRISPR/Cas in potato genome editing.

Keywords: conventional breeding, transgenics, ZFN, TALEN, CRISPR/Cas, biotic and abiotic stress resistance, tuber quality and browning, starch content

1. Introduction

Potato (*Solanum tuberosum*), a member of the Solanaceae family, is at fourth position among the most important staple food crops in the globe succeeded by rice and wheat in terms of human consumption [1]. By 2050, the estimated world population will be 9.7 billion and potato will have a vital role in the future in securing food resources for the human beings [2]. From breeding point of view, potato has both morphological and genetical advantages and disadvantages. As potato is asexually propagated, it is exempted from the need to be bred from true seeds to produce homogenous plants. Being tetraploid makes it extremely difficult for desired features to be passed on to offspring in the future [3]. Moreover, potato research is complicated and time-taking process as it is highly heterozygous and polyploid crop, which cause significant hinderance while utilizing standard breeding approaches, ultimately making use of genome editing indispensable [4, 5]. At present, research is being conducted for improvement of specific desirable traits in several crops through genome editing (GE) technologies. Different genome editing approaches are considered as revolutionary for crop improvement as gene knockout and insertion/deletion mutagenesis are possible through genome editing technologies [6]. Crop improvement using genome editing technologies come up with multiple possibilities, such as:

Alteration in only one or two nucleotides among billions obtained from genome of living cells

- Alteration in whole alleles
- Insertion of a non-existing gene in the targeted region of the genome.
- Due to high accuracy of genome editing tools, these are preferred over traditional plant breeding approaches and standard genetic engineering approaches. Along with enhancing the nutritional profile of crops, gene editing tools can effectively be used for incorporating biotic and abiotic stress resistance/tolerance and can develop crop varieties that survive well in adverse climatic conditions such as arid climate [7]. Therefore, gene editing tools can be effectively used for securing the world's food supply.

Genome editing tools permits double-stranded breaks (DSBs) at particular genomic locations and recovers them using inherent DNA repair mechanisms, such as non-homologous end joining (NHEJ) or homologous recombination (HR). Earlier, this process was assisted by using protein-guided nucleases including zinc finger nucleases (ZFNs) and transcription activator-like effector nucleases (TALENs). But lately, the RNA-guided nuclease known as Clustered Regularly Interspaced Short

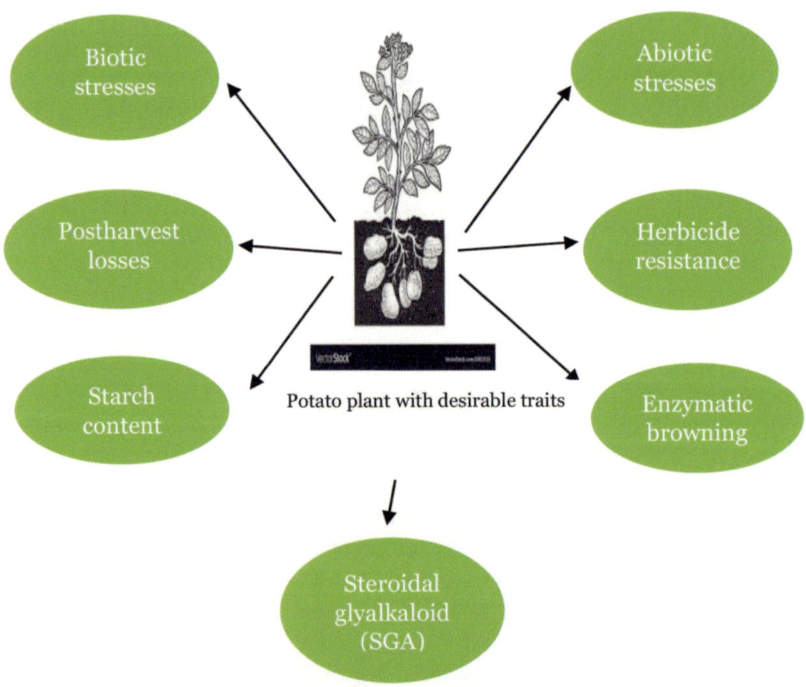

Figure 1.
Advances in potato improvement using clustered regularly interspaced short palindromic repeats (CRISPR)-based genome editing.

Target gene	Delivery method	Editing agent	Phenotype	Reference
Starch quality				
GBSSI	Protoplast	Cas 9	Decrease in amylose content	[9]
GBSSI	*A. tumefaciens*	Cas 9, PmCDA1-CBE	N/A	[10]
GBSSI	*A. tumefaciens*	Cas 9	Decrease in amylose content	[11]
SBE1, SBE2	Protoplast	Cas 9	Decrease in amylose content	[12]
Browning				
PPO2	Protoplast	Cas 9	Reduced browning	[13]
PPO2	Protoplast, *A. tumefaciens*	Cas 9	Reduced browning	[13]
Steroidal glyalkaloid (SGA)				
SSR2	*A. tumefaciens*	Cas 9	66% of WT tuber	[14]
Biotic stress tolerance				
eIF4E1	Protoplast	Cas 9	Partial resistance to PVY	[15]
RNase III	*A. tumefaciens*	Cas 13	Improved resistance to sweet potato virus disease	[16]
Abiotic stress tolerance				
MYB44	*A. tumefaciens*	Cas 9	N/A	[17]
Herbicide tolerance				
ALS1	*A. tumefaciens*	Prime Editor 2	Improved herbicide resistance	[18]

Table 1.
CRISPR/Cas applications in potato genome editing.

Palindromic Repeats (CRISPR)—CRISPR associated—has captured the researchers' attention (Cas) [8]. Both ZFNs and TALENs requires more time and technical expertise than CRISPR/Cas. Numerous significantly important CRISPR/Cas studies have been conducted in potato, here we are featuring some selected achievements of CRISPR-based editing in potato (**Figure 1** and **Table 1**). This chapter provides the current status of CRISPR/Cas, future perspectives, and challenges in potato.

2. Key genome editing approaches

2.1 ZFNs (Zinc-finger nucleases)

ZFNs were first created by Kim and colleagues [19]. ZFNs develop as a result of the interaction between DNA-binding and DNA-cleaving domains [20]. These synthetic proteins were created based on the fact that the natural restriction enzyme FokI exhibits unique binding and cleaving capabilities [21]. The binding domain consists of eukaryotic transcription factors and 3–6 zinc finger repeats which recognize between 9 and 18 base pairs [22], whereas the DNA-cleavage domain has a FokI restriction enzyme DNA cleavage. To cleave the DNA, the zinc finger domain fused at its C-termini via a peptide linker to the FokI of the cleavage domain. This cleavage domain must dimerize in order to cut DNA. For two cleavage domains to dimerize, the two different ZFNs with their C-termini must bind to opposing DNA strands. If zinc finger, domains are unable to target their specific site within the genome, then off-target cleavage occurs. This off-target cleavage may result in significant double-strand breaks that stop the repair process, which results in cell death or chromosome rearrangements. Further, this leads to the random integration of donor DNA [23]. These domains can be customized to target desired DNA sequences within the complex genomes. By utilizing endogenous DNA repair mechanisms, ZFNs can be used for the precise genome alteration of higher organisms. The efficiency of a given ZFN pair depends on its binding affinity and sequence specificity, both of which impact long-term stability and on-target modification [24]. It is a time taking process to design successful ZFNs. Moreover, ZFNs have poor targeting density which leads to off-targets.

2.2 TALENs

Transcription activator-like effector nucleases (TALENs) are synthetic restriction enzymes that merge the FokI nucleases with the DNA-binding domain of TALEs. TALEs are the natural lethal proteins derived from the Xanthomonas bacteria that bind with specifically targeted DNA sites via a DNA-binding domain in the centre to activate host gene expression. This central domain uses a unique DNA-binding mechanism that involves one-to-one correspondence between an individual repeat and a single nucleotide. Each TALE array has fifteen to nineteen individual tandem repeats and is highly conserved that varies only at amino acid positions twelve and thirteen, known as repeat variable di-residue (RVD) [25]. The RVDs determine the binding specificity of DNA with the TALE array base on one-to-one correspondence along with the base as specified RVD. There are twenty-five types of natural RVDs with good binding specificity, among which NI (Asn-IIe), HD (His-Asp), NH (Asn-Gly), specific for identifying adenine (A), cytosine (C), guanine (G) and thymine (T) [25, 26]. The binding specificity of TALENs is affected by cell type, target sites, duration of effect and delivery system used. The coding for the recognition of the TALENs sequence is comparatively easy, which is an advantage with respect to its targeting density when compared with ZFNs [27]. It allows genome editing with more specificity and low cytotoxicity.

2.3 CRISPR/Cas

CRISPR/Cas is a potent tool for genetic manipulation that precisely mutates specific DNA sequences [28]. There are different CRISPR/Cas systems that differentiate on the basis of the nuclease effector used. This diversity divided the CRISPR/Cas systems into two classes (single/multi-subunit) based on the structure and into six types and 27 subtypes on the basis of the functions of each subunit [29, 30]. It has two components: the Cas9 nuclease and sgRNA. The Cas9 nuclease has two lobes viz., REC for recognition and NUC with nuclease activity. The non-coding sgRNA is complementary to the target sequence (protospacer) of 20 base pairs. This protospacer has a three-base sequence at its 3′ end called PAM (Protospacer Adjacent Motif), recognized by the NUC lobe to produce double-strand breaks (DSB). The DSBs are created when sgRNA/Cas complex binds with the target DNA sequence. Further, these DSBs are repaired by NHEJ (Non-homologous end-joining) and HDR (Homologous directed repair) mechanisms [31].

3. Genome editing in potato

Introduction of genes related to economically important traits from wild species is very difficult and time intensive assignment in potato (*Solanum tuberosum*) is as it is a heterozygous polyploid crop. Due to this, traditional breeding approaches cease to work properly when numerous traits and/or new traits which are not available in gene pool need to be introgressed for crop advancement. As genome sequencing data of potato is accessible from public database and standard genetic modification and regeneration protocols are also obtainable, which made potato a prime candidate for gene editing. Thus, gene editing approaches can be used for potato improvement by improving the production and quality characteristics of the crop without affecting maximum allele combination in commercial cultivars [32–36]. In tetraploid potato, the earliest successful utilization of TALENs were reported for knocking out alleles related to sterol side chain reductase-2 (StSSR2) [37], which are responsible for production of anti-nutritional sterol glycoalkaloid (SGA) [38, 39]. In 2015, it was reported that genome editing approaches such as TALENs [40] and CRISPR/Cas [41] can be utilized to precisely alter the potato genome. CRISPR/Cas9 mediated targeted mutation of StIAA2 gene, encodes for Aux/IAA protein, resulted in homozygous mono and biallelic mutations in the first generation of modified plants [41].

4. Trait improvement in potato using CRISPR/Cas technology

4.1 Disease resistance

Different biotic stresses are the root cause behind major obstruction in potato production, which ultimately results in enormous loss to farmers. Even now, several scientists are trying to make a key advancement in development of biotic stress resistant potatoes with the help of genome editing approaches. CRISPR/Cas technology of genome editing is considered as an effective substitute for advancing potato breeding programmes.

Phytophthora infestans cause late blight disease of potato, which is the key hinderance in enhancing potato production [42]. Therefore, many researchers are emphasizing on the development of late blight resistant potato cultivars, which can be achieved through knocking out or deleting susceptible genes (*S*-genes) of late blight [43]. At present, fungicidal sprays and breeding methods are used for the management of late blight in potato. At a recent time, late blight resistance in potato was commenced through knocking out of susceptible genes StDMR6-1 and StCHL1 [44] and Caffeoyl-CoA O-methyltransferase (StCCoAOMT) [45]. Extracellular receptor protein ELR (elicitin response) is found in *Solanum microdontum*, a wild potato species. Du and co-workers [46] documented utilization of ELR protein for identification of an elicitin which is considered as an extremely conserved in *Phytophthora* species contributing a wide range durable resistance against late blight pathogen.

For production of potato virus Y (PVY) resistant potatoes, four viral genes, P3, CI, Nib, and CP were targeted through CRISPR/Cas approach. Cas13a protein was utilized to confer resistant against three strains of RNA virus, PVY [47]. Makhotenko and co-workers [48] demonstrated efficient use of host genes (eukaryotic translation initiation factor eIF4E and coilin) for development of PVY resistant potatoes.

4.2 Resistance/tolerance to abiotic stresses

Abiotic stresses include high and low temperatures, drought and salinity cause major constraints in potato production; still limited research work is documented until now. In 2017, Zhou et al. [17] modified MYB transcription factor gene (StMYB44), responsible for restraining StPHO1 gene expression which negatively control phosphate transport activity in potato, with a proficiency rate of 85%. In 2020, Tiwari et al. [49] documented the utilization of gene editing approach CRISPR/Cas for modification of nitrogen metabolism genes to enhance the nitrogen use efficiency of potato plants.

4.3 Herbicide resistance

Butler and co-workers [50] developed a ss gemini virus-based DNA replicon (GVR) as a vector for transfer of TALEN genes and a mutated fragment of ALS1 gene. TALEN genes were transferred to target Acetolactate synthase1 (ALS1) gene in potato while the mutated ALS1gene fragment verified tolerance for numerous categories of ALS-inhibiting herbicides.

4.4 Post harvest

Potato is harvested annually which makes it imperative for producers and sellers to store these tubers in cold stores. Cold storage reduces sprouting and extend shelf life. But during storage in cold storage houses the sucrose in tubers is reduced to sugars which react with amino acids on heating and cause the processed potato products to turn brown and bitter. Moreover, the acrylamide level also goes up. This deterioration in organoleptic properties and enhancement in acrylamide levels negatively affects potato tuber quality thereby leading to a reduced public demand. Accretion of reducing sugars through cold induced sweetening is swayed by several metabolic processes *viz* starch synthesis, degradation, glycolysis, hexogenesis and mitochondrial respiration. A family of ubiquitous enzymes termed invertases degrades starch into sucrose and fructose [51]. These invertases have been sub-localized to cell wall, vacuole and cytoplasm. Vacuole localized invertases (*VInv)* play a significant role in cold induced sugar production. Therefore, these have been targeted most to prevent cold induced sweetening. Silencing *VInv* gene has been accomplished mostly by RNAi [52], TALEN [23] and CRISPR/Cas technologies [53]. Significant reduction in sugar levels were observed in the transgenics developed via these technologies.

4.5 Starch

Potato is a valuable source of starch. Potato starch is used for both culinary and industrial purposes. It is often made up of amylose (20–30%) and amylopectin (70–80%) [54]. The amylose/amylopectin ratio influences the characteristics of both dietary and industrial starch [55]. Amylose-free starch has better freeze-thaw stability and is thus a key component in the manufacturing of frozen foods. Amylopectin has good binding properties and is therefore employed in the paper and glue industries [56]. Therefore, to increase amylopectin content; earlier the genetic engineering efforts primarily aim to produce amylose-free (waxy) potatoes. This might be accomplished by targeting specific genes involved in starch production [57]. Potatoes that produce amylopectin starch may be created by knocking out or silencing the GBSS gene. Various methods have been used to accomplish development of amylopectin rich cultivars viz., radiation-induced mutagenesis, antisense technologies like RNAi, TALENs, and, more recently, CRISPR/Cas [58]. Amylose-free or amylose-reduced lines of sweet potato cultivars have also been developed utilizing CRISPR/CAS genome editing system [59].

Nowadays main focus of researchers has been on altering starch properties to reduce its harmful effects on humans prone to obesity and diabetes. Potatoes with less digestible starch i.e., resistant starch has been engineered by RNAi mediated silencing of genes encoding for starch binding enzymes (SBE). Essentially identical results have been achieved by different means of reducing expression of both SBE isoforms (SBE1 and SBE2): antisense RNA [60], expression of single-domain SBE-specific camelid antibodies [61] and CRISPR/Cas [12, 62]. Though genome editing technologies proved useful in lowering or boosting starch content in potatoes but transient application of vectors harboring inserted DNA fragments lead to the unwanted insertions of vector genome into targeted regions [63].

4.6 Steroidal glyalkaloid (SGA)

Solanaceous plants produce secondary metabolites called glycoalkaloids. More than 80 steroidal glyalkaloids have been identified in different potato species; the two most common ones in cultivated potatoes are α-solanine and α-chaconine [64]. Higher levels of steroidal glyalkaloids have neurotoxic and anti-nutritional effects on human [65]. Therefore, the amount of steroidal glyalkaloids in tubers meant for eating should not exceed the threshold value of 200 mg/kg fresh weight. Nonetheless, significant efforts have been made to reduce SGA content. RNAi suppression of the SGA biosynthetic genes sterol side chain reductase 2 (StSSR2) [26] and GLYCOALKALOID METABOLISM 1 (GAME1) [66] resulted in decreased SGA content. One study on the CRISPR/Cas system's use for controlling SGA levels in potatoes looked at knocking down the gene encoding steroid 16-hydroxylase (St16DOX), which appears in the genome in a single copy [67]. Two mutants showed positive results for no detectable levels of solanine or chaconine but the plants as a whole were not regenerated so success for CRISPR/Cas-mediated deletion of St16DOX was partial and not all attempts at SGA reduction in potato were equally successful [67]. CRISPR/Cas9-editing of StSSR2 was used to manipulate the SGA level in the cv. Atlantic with significant success [14]. But on the other hand, SGAs are the plant's first line of defense against viruses and herbivores and participate in ecological interactions with microorganisms therefore eliminating SGAs would be damaging to the plant.

4.7 Enzymatic browning

High tuber quality is a desirable characteristic in the potato processing business and among consumers. Enzymatic browning (EB) causes nutritional quality loss and alters the flavor and texture of tubers. Mechanical injury caused when tubers are cut, sliced or peeled disrupts subcellular compartmentation, resulting in the release of amyloplast-localized PPOs and vacuole-localized phenolic compounds [68]. PPOs catalyze the oxidation of monophenols and/or o-diphenols to o-quinones in the presence of oxygen, which then polymerize and form complexes with proteins, resulting in brown pigment build-up [69]. This enzyme is coded by a multi-gene family, and five PPO genes in potato have been recognized thus far, with numerous allele variants for each gene [68]. Though enzymatic browning is a quantitative characteristic, PPOs are frequently used as target genes to minimize browning in tubers because the PPO loci have a strong correlation with the QTL associated with browning of tubers [70]. Because PPOs are involved in numerous physiological functions as well as defense against diseases and pests, knocking out all PPO genes would almost certainly be fatal to plants. As a result, specific gene targeting or allele silencing is necessary [68]. Substantial progress has been made in this direction. StPPO2 gene was effectively modified in the cv. Desiree via CRISPR/Cas9 [13]. Two different types of edited lines were created by the four-allele StPPO2 edition: those with 111-bp non-frameshift alterations and those with frameshift mutations in the coding region that may have affected the activity of the enzyme after translation. No matter the mutation type, all lines had lower PPO expression levels and were less susceptible to EB than the wild type [13]. GM potato varieties (Innate® potato) with decreased enzymatic browning (EB) and minimal acrylamide generation have been successfully developed and are under cultivation [71].

5. Challenges in potato gene editing

Potato is an asexually propagated, heterozygous and polyploid crop, which causes multiple obstacles in utilization of genome editing approaches. Major obstacles are: complications in target designing for gene editing and acquiring homozygous mutants with mutation of all targeted gene(s), which make it essential to screen a vast number of transformed plants to identify and propagate mutated lines carrying multiple alleles. Additionally, not each and every potato variety is responsive to transformation. Protoplast mediated transformation and regeneration of plants from leaf as an explant can also cause somaclonal variation, ultimately negatively affecting plant development [8].

To understand complicated economically important traits, breeders are making efforts for development of diploid lines of potato. Self-incompatibility (SI) causes the significant obstruction in development of potato inbred lines, as SI impede in fixing of gene edits and selection of progeny by segregating out the inserted gene. Ye and co-workers [72] used CRISPR/Cas9 approach for production of diploid self-compatible lines of potato by knocking out the Stylar ribonuclease gene (*S-RNase*), responsible for self-incompatibility in potato. Yet a number of diploid, SI potato lines shows recalcitrant response to transformation through traditional *Agrobacterium tumefaciens* [73]. To avoid this, Butler and co-workers [73] used *A. rhizogenes* for speedy production of stable mutants within hairy root clones in potato genotypes which were already showing recalcitrant nature to *A. tumefaciens*. But in this method, analysis of hairy root clones was the major drawback. Genome editing approach CRISPR/Cas9 was implemented for targeting *phytoene desaturase* (*StPDS*) gene in potato. This gene is found in hairy root clones of potato. This method resulted in 64–98% mutation in transformed hairy root clones.

Production of off-target mutants in non-target gene(s) in potato while genome editing is an additional area to be studied. Off-target mutants cause unacceptable alterations in plants and also it complicates the procedure of mutant analysis. Synthetic proofreading Cas9 variants [74], test of sgRNA activity and developing good design [75] are the approaches used to decrease or even remove these off-target mutants.

Also, it is important to produce transgene or foreign gene free potato. To be adopted by consumers and accepted by policy makers, gene edited (GE) crops should not contain any traces of foreign DNA [76].

6. Conclusion

Genome editing can perform a key role in potato improvement by enhancing the tuber starch content, developing resistance/tolerance against biotic/abiotic stresses and reducing antinutritional factors and toxic compounds. As genome editing approaches are highly effective and accurate, this escalates the possibilities of improvement of other economically valuable plant characteristics also. In case of potato, it requires much efforts and skills for advancement of plant characters governed by multiple genes as compared to those governed by single gene due to heterozygous polyploid nature and asexual reproduction of potato plants. In spite of all these bottlenecks, significant success has been obtained in potato for few plant characteristics and many of them using gene editing approaches (gene knockout or addition/deletion of gene). Researchers have reported multiplexing of SpCas9 along with protoplast mediated transformation as an ideal choice for potato improvement. Over and above, it is important to aware consumers about difference between genome edited (GE) crops or crop varieties and genetically modified (GM) crops or crop varieties. Efforts are being made to keep genetically edited and genetically modified crops separate, which is important for advancement of gene editing approaches and their success. Collectively, genome editing through CRISPR/Cas is an efficient and precise next generation approach for rapid potato breeding to obtain sustainable crop yield.

Author details

Deepa Beniwal[1*], Shivani Chauhan[2] and Harnoor Kaur Dhillon[2]

1 ARS, SKRAU, Bikaner, Rajasthan, India

2 Department of Vegetable Science, Punjab Agricultural University, Ludhiana, Punjab, India

*Address all correspondence to: deepabeniwal24@gmail.com

References

[1] Tiwari JK, Buckseth T, Challam C, Zinta R, Bhatia N, Dalamu D, et al. CRISPR/Cas genome editing in potato: Current status and future perspectives. Frontiers in Genetics. 2022;**13**:827808. DOI: 10.3389/fgene.2022.827808

[2] Jacobs MMJ, Smulders MJM, Van den Berg RG, Vosman B. What's in a name; genetic structure in *Solanum* section Petota studied using population-genetic tools. BMC Evolutionary Biology. 2011;**11**:42. DOI: 10.1186/1471-2148-11-42

[3] Kennedy GG. In: Romeis J, Shelton AM, Kennedy GG, editors. Integration of Insect-Resistant Genetically Modified Crops within IPM Programs. Amsterdam: Springer; 2008. pp. 1-26

[4] Andersson M, Turesson H, Olsson N, Falt A, Ohlsson P, Gonzalez MN, et al. Genome editing in potato via CRISPR-Cas9 ribonucleoprotein delivery. Physiologia Plantarum. 2018;**164**:378-384. DOI: 10.1111/ppl.12731

[5] Muthoni J, Kabira J, Shimelis H, Melis R. Tetrasomic inheritance in cultivated potato and implications in conventional breeding. Australian Journal of Crop Science. 2015;**93**:185-190. DOI: 10.3316/informit.075206802413826

[6] Hameed A, Zaidi SS, Shakir S, Mansoor S. Applications of new breeding technologies for potato improvement. Frontiers in Plant Science. 2018;**9**:925. DOI: 10.3389/fpls.2018.00925

[7] Voytas DF, Gao C. Precision genome engineering and agriculture: Opportunities and regulatory challenges. PLoS Biology. 2014;**12**:e1001877. DOI: 10.1371/journal.pbio.1001877

[8] Nadakuduti SS, Buell CR, Voytas DF, Starker CG, Douches DS. Genome editing for crop improvement-applications in clonally propagated polyploids with a focus on potato (*Solanum tuberosum* L.). Frontiers in Plant Science. 2018;**9**:1607. DOI: 10.3389/fpls.2018.01607

[9] Johansen IE, Liu Y, Jorgensen B, et al. High efficacy full allelic CRISPR/Cas9 gene editing in tetraploid potato. Scientific Reports. 2019;**9**:17715. DOI: 10.1038/s41598-019-54126-w

[10] Veillet F, Kermarrec MP, Chauvin L, Chauvin JE, Nogue F. CRISPR-induced indels and base editing using the *Staphylococcus aureus* Cas9 in potato. PLoS One. 2020;**15**(8):e0235942. DOI: 10.371/journal.pone.0235942

[11] Toinga-Villafuerte S, Vales MI, Awika JM, Rathore KS. CRISPR/Cas9-mediated mutagenesis of the granule-bound starch synthase gene in the potato variety Yukon Gold to obtain amylose-free starch in tubers. International Journal of Molecular Sciences. 2022;**23**(9):4640. DOI: 10.3390/ijms23094640

[12] Zhao X, Jayarathna S, Turesson H, et al. Amylose starch with no detectable branching developed through DNA-free CRISPR-Cas9 mediated mutagenesis of two starch branching enzymes in potato. Scientific Reports. 2021;**11**:4311. DOI: 10.1038/s41598-021-83462-z

[13] Gonzalez MN, Massa GA, Andersson M, Turesson H, Olsson N, Falt AS, et al. Reduced enzymatic browning in potato tubers by specific editing of a polyphenol oxidase gene via ribonucleoprotein complexes delivery of the CRISPR/Cas9 system. Frontiers in Plant Science. 2020;**10**:1649. DOI: 10.3389/fpls.2019.01649

[14] Zheng Z, Ye G, Zhou Y, Pu X, Su W, Wang J. Editing sterol side chain reductase 2 gene (StSSR2) via CRISPR/Cas9 reduces the total steroidal glycoalkaloids in potato. All Life. 2021;**14**(1):401-413. DOI: 10.1080/26895293.2021.1925358

[15] Lucioli A, Tavazza R, Baima S, Fatyol K, Burgyan J, Tavazza M. CRISPR-Cas9 targeting of the eIF4E1 gene extends the potato virus Y resistance in spectrum of the *Solanum tuberosum* L. cv. Desiree. Frontiers in Microbiology. 2022;**13**:873930. DOI: 10.3389/fmicb.2022.873930

[16] Yu Y, Pan Z, Wang X, Bian X, Wang W, Liang Q, et al. Targeting of SPCSV-RN*ase3* vis CRISPR-Cas13 confers resistance against sweet potato virus disease. Molecular Plant Pathology. 2021;**23**(1):104-117. DOI: 10.1111/mpp.13146

[17] Zhou X, Zha M, Huang J, Li L, Imran M, Zhang C. StMYB44 negatively regulates phosphate transport by supressing expression of *PHOAPHATE1* in potato. Journal of Experimental Botany. 2017;**68**(5):1265-1281. DOI: 10.1093/jxb/erx026

[18] Perroud PF, Guyon-Debast A, Veillet F, Kermarrec MP, Chauvin L, Chauvin JE, et al. Prime editing in the model plant *Physcomitrium patens* and its potential in the tetraploid potato. Plant Science. 2022;**316**:111162. DOI: 10.1016/j.plantsci.2021.111162

[19] Kim YG, Cha J, Chandrasegaran S. Hybrid restriction enzymes: Zinc finger fusions to *Fok*I cleavage domain. Proceedings of the National Academy of Sciences of the United States of America. 1996;**93**:1156-1160. DOI: 10.1073/pnas.93.3.1156

[20] Cathomen T, Joung JK. Zinc-finger nucleases: The next generation emerges. Molecular Therapy. 2008;**16**:1200-1207. DOI: 10.1038/mt.2008.114

[21] Li L, Wu LP, Chandrasegaran S. Functional domains in *Fok*I restriction endonuclease. Proceedings of the National Academy of Sciences of the United States of America. 1992;**89**:4275-4279. DOI: 10.1073/pnas.89.10.4275

[22] Handel EM, Cathomen T. Zinc-finger nuclease-based genome surgery: It's all about specificity. Current Gene Therapy. 2011;**11**:28-37. DOI: 10.2174/156652311794520120

[23] Carroll D. Genome engineering with zinc-finger nucleases. Genetics. 2011;**188**:773-782. DOI: 10.1534/genetics.111.131433

[24] Urnov FD, Miller JC, Lee YL, Beausejour CM, Rock JM, Augustus S, et al. Highly efficient endogenous human gene correction using designed zinc-finger nucleases. Nature. 2005;**435**:646-651. DOI: 10.1038/nature03556

[25] Richter A, Streubel J, Boch J. TALEN effector DNA-binding principles and specificity. In: Kühn R, Wurst W, Wefers B, editors. TALENs. Vol. 25. New York, NY, USA: Springer; 2016. pp. 1338-1339. DOI: 10.1007/978-1-4939-2932-0_2

[26] Bloom K, Mussolino C, Arbuthnot P. Transcription Activator-Like Effector (TALE) nucleases and repressor TALEs for antiviral gene therapy. Current Stem Cell Reports. 2015;**1**:1-8. DOI: 10.1007/s40778-014-0008-7

[27] Mussolino C, Alzubi J, Fine EJ, Morbitzer R, Cradick TJ, Lahaye T, et al. TALENs facilitate targeted genome editing in human cells with high specificity and low cytotoxicity. Nucleic Acids Research. 2014;**42**:6762-6773. DOI: 10.1093/nar/gku305

[28] Charlesworth CT, Deshpande PS, Dever DP, Camarena J, Lemgart VT, Cromer MK, et al. Identification of pre-existing adaptive immunity to Cas9 proteins in humans. Nature Medicine. 2019;**25**:249-254. DOI: 10.1038/s41591-018-0326-x

[29] Koonin EV, Makarova KS. Origins and evolution of CRISPR-Cas systems. Philosophical Transactions of the Royal Society B Biological Sciences. 2019:374. DOI: 10.1098/rstb.2018.0087

[30] Barman A, Deb B, Chakraborty S. A glance at genome editing with CRISPR–Cas9 technology. Current Genetics. 2019;**66**:447-462. DOI: 10.1007/s00294-019-01040-3

[31] Liu M, Rehman S, Tang X, Gu K, Fan Q, Chen D, et al. Methodologies for improving HDR efficiency. Frontiers in Genetics. 2019;**9**:691-699. DOI: 10.3389/fgene.2018.00691

[32] Diretto G, Tavazza R, Welsch R, Pizzichini D, Mourgues F, Papacchioli V, et al. Metabolic engineering of potato tuber carotenoids through tuber specific silencing of lycopene epsilon cyclase. BMC Plant Biology. 2006;**6**:13. DOI: 10.1186/1471-2229-6-13

[33] Chi M, Bhagwat B, Lane W, Tang G, Su Y, Sun R, et al. Reduced polyphenol oxidase gene expression and enzymatic browning in potato (*Solanum tuberosum* L.) with artificial microRNAs. BMC Plant Biology. 2014;**14**:62. DOI: 10.1186/1471-2229-14-62

[34] Clasen BM, Stoddard TJ, Luo S, Demorest ZL, Li J, Cedrone F, et al. Improving cold storage and processing traits in potato through targeted gene knockout. Plant Biotechnology. 2016;**14**:169-176. DOI: 10.1111/pbi.12370

[35] Sun Y, Jiao G, Liu Z, Zhang X, Li J, Guo X, et al. Generation of high amylose rice through CRISPR/Cas9-mediated targeted mutagenesis of starch branching enzymes. Frontiers in Plant Science. 2017;**8**:298. DOI: 10.3389/fpls.2017.00298

[36] McCue KF, Breksa A, Vilches A, Belknap WR. Modification of potato steroidal glycoalkaloids with silencing RNA constructs. American Journal of Potato Research. 2018;**95**:317. DOI: 10.1007/s12230-018-9658-9

[37] Sawai S, Ohyama K, Yasumoto S, Seki H, Sakuma T, Yamamoto T, et al. Sterol side chain reductase 2 is a key enzyme in the biosynthesis of cholesterol, the common precursor of toxic steroidal glycoalkaloids in potato. The Plant Cell. 2014;**26**:3763-3774. DOI: 10.1105/tpc.114.130096

[38] Itkin M, Rogachev I, Alkan N, Rosenberg T, Malitsky S, Masini L, et al. Glycoalkaloid metabolism is required for steroidal alkaloid glycosylation and prevention of phytotoxicity in tomato. The Plant Cell. 2011;**23**:4507-4525. DOI: 0.1105/tpc.111.088732

[39] Itkin M, Heinig U, Tzfadia O, Bhide AJ, Shinde B, Cardenas PD, et al. Biosynthesis of antinutritional alkaloids in solanaceous crops is mediated by clustered genes. Science. 2013;**341**:175-179. DOI: 10.1126/science. 1240230

[40] Nicolia A, Proux-Wéra E, Åhman I, Onkokesung N, Andersson M, Andreasson E, et al. Targeted gene mutation in tetraploid potato through transient TALEN expression in protoplasts. Journal of Biotechnology. 2015;**204**:17-24. DOI: 10.1016/j.jbiotec.2015.03.021

[41] Wang S, Zhang S, Wang W, et al. Efficient targeted mutagenesis in potato by the CRISPR/Cas9 system. Plant Cell Reports. 2015;**34**:1473-1476. DOI: 10.1007/s00299-015-1816-7

[42] Fisher MC, Henk DA, Briggs CJ, Brownstein JS, Madoff LC, McCraw SL, et al. Emerging fungal threats to animal, plant and ecosystem health. Nature. 2012;**484**:186-194. DOI: 10.1038/nature10947

[43] Schaart JG, van de Wiel CC, Lotz LA, Smulders MJ. Opportunities for products of new plant breeding techniques. Trends in Plant Science. 2016;**21**(5):438-449. DOI: DOI.org/10.1016/j.tplants.2015.11.006

[44] Kieu NP, Lenman M, Wang ES, Petersen BL, Andreasson E. Mutations introduced in susceptibility genes through CRISPR/Cas9 genome editing confer increased late blight resistance in potatoes. Scientific Reports. 2021;**11**(1):4487. DOI: 10.1038/s41598-021-83972-w

[45] Hegde N, Joshi S, Soni N, Kushalappa AC. The Caffeoyl-CoA O-Methyltransferase gene SNP replacement in Russet burbank potato variety enhances late blight resistance through cell wall reinforcement. Plant Cell Reports. 2021;**40**(1):237-254. DOI: 10.1007/s00299-020-02629-6

[46] Du J, Verzaux E, Chaparro-Garcia A, Bijsterbosch G, Keizer LCP, Zhou J, et al. Elicitin recognition confers enhanced resistance to *Phytophthora infestans* in potato. Nature Plants. 2015;**1**:15034. DOI: 10.1038/NPLANTS.2015.34

[47] Zhan X, Zhang F, Zhong Z, Chen R, Wang Y, Chang L, et al. Generation of virus-resistant potato plants by RNA genome targeting. Plant Biotechnology. 2019;**17**:1814-1822. DOI: 10.1111/pbi.13102

[48] Makhotenko AV, Khromov AV, Snigir EA, Makarova SS, Makarova VV, Suprunova TP, et al. Functional analysis of coilin in virus resistance and stress tolerance of potato *Solanum tuberosum* using CRISPR-Cas9 editing. Doklady Biochemistry and Biophysics. 2019;**484**:88-91. DOI: 10.1134/s1607672919010241

[49] Tiwari JK, Buckseth T, Singh RK, Kumar M, Kant S. Prospects of improving nitrogen use efficiency in potato: Lessons from transgenics to genome editing strategies in plants. Frontiers in Plant Science. 2020;**11**:597481. DOI: 10.3389/fpls.2020.597481

[50] Butler NM, Baltes NJ, Voytas DF, Douches DS. Gemini virus-mediated genome editing in potato (*Solanum tuberosum* L.) using sequence-specific nucleases. Frontiers in Plant Science. 2016;**7**:1045. DOI: 10.3389/fpls.2016.01045

[51] Drafehn AM, Meller S, Li L, Gebhardt C. Natural diversity of potato (*Solanum tuberosum*) invertases. BMC Plant Biology. 2010;**10**:1-15. DOI: 10.1186/1471-2229-10-271

[52] Zhu X, Gong H, He Q, Zeng Z, Busse JS, Jin W, et al. Silencing of vacuolar invertase and asparagine synthetase genes and its impact on acrylamide formation of fried potato products. Plant Biotechnology Journal. 2016;**14**:709-718. DOI: 10.1111/PBI.12421

[53] Shumbe L, Visse M, Soares E, Smit I, Dupuis B, Vanderschuren H. Differential DNA methylation in the Vinv promoter region controls cold induced sweetening in potato. bioRxiv. 2020:1-32. DOI: 10.1101/2020.04.26.062562

[54] Fajardo D, Haynes KG, Jansky S. Starch characteristics of modern and heirloom potato cultivars. American Journal of Potato Research. 2013;**90**:460-469. DOI: 10.1007/S12230-013-9320-5

[55] Karlsson ME, Leeman AM, Bjorck IME, Eliasson AC. Some physical and nutritional characteristics of genetically modified potatoes varying in amylose/amylopectin ratios. Food Chemistry. 2007;**100**:136-146. DOI: 10.1016/j.foodchem.2005.09.032

[56] Slattery CJ, Kavakli IH, Okita TW. Engineering starch for increased quantity and quality. Trends in Plant Science. 2000;**5**:291-298. DOI: 10.1016/S1360-1385(00)01657-5

[57] Andersson M, Turesson H, Nicolia A. Efficient targeted multiallelic mutagenesis in tetraploid potato (*Solanum tuberosum*) by transient CRISPR-Cas9 expression in protoplasts. Plant Cell Reports. 2017;**36**:117-128. DOI: 10.1007/s00299-016-2062-3

[58] Kusano H, Ohnuma M, Mutsuro-Aoki H, Asahi T, Ichinosawa D, Onodera H, et al. Establishment of a modified CRISPR/Cas9 system with increased mutagenesis frequency using the translational enhancer dMac3 and multiple guide RNAs in potato. Scientific Reports. 2018;**81**(8):13753. DOI: 10.1038/s41598-018-32049-2

[59] Wang H, Wu Y, Zhang Y, Yang J, Fan W, Zhang H, et al. CRISPR/Cas9-based mutagenesis of starch biosynthetic genes in sweet potato (*Ipomoea batatas*) for the improvement of starch quality. International Journal of Molecular Sciences. 2019;**20**:4702. DOI: 10.3390/ijms20194702

[60] Schwall GP, Safford R, Westcott RJ, Jeffcoat R, Tayal A, Shi YC, et al. Production of very-high-amylose potato starch by inhibition of SBE A and B. Nature Biotechnology. 2000;**18**(5):555-554. DOI: 10.1038/75427

[61] Jobling SA, Jarman C, Teh MM, Holmberg N, Blake C, Verhoeyen ME. Immunomodulation of enzyme function in plants by single-domain antibody fragments. Nature Biotechnology. 2003;**21**:77-80

[62] Tuncel A, Corbin KR, Ahn JJ, Harris S, Hawkins E, Smedley MA, et al. Cas9-mediated mutagenesis of potato starch-branching enzymes generates a range of tuber starch phenotypes. Plant Biotechnology Journal. 2019;**17**:2259-2271. DOI: 10.1111/pbi.13137

[63] Veillet F, Chauvin L, Kermarrec MP, Sevestre F, Merrer M, Terret Z, et al. The *Solanum tuberosum* GBSSI gene: A target for assessing gene and base editing in tetraploid potato. Plant Cell Reports. 2019;**38**:1065-1080. DOI: 10.1007/s00299-019- 02426-w

[64] Ginzberg I, Tokuhisa JG, Veilleux RE. Potato steroidal glycoalkaloids: Biosynthesis and genetic manipulation. Potato Research. 2009;**52**:1-15. DOI: 10.1007/s11540-008-9103-4

[65] Delbrouck JA, Desgagné M, Comeau C, Bouarab K, Malouin F, Boudreault PL. The therapeutic value of solanum steroidal (glyco) alkaloids: A 10-year comprehensive review. Molecules. 2023;**28**(13):4957. DOI: 10.3390/molecules28134957

[66] Cardenas PD, Sonawane PD, Pollier J, Bossche RV, Dewangan V, Weithorn E, et al. GAME9 regulates the biosynthesis of steroidal alkaloids and upstream isoprenoids in the plant mevalonate pathway. Nature Communications. 2016;**71**(7):10654. DOI: 10.1038/ncomms10654

[67] Nakayasu M et al. Generation of α-solanine-free hairy roots of potato by CRISPR/Cas9 mediated genome editing of the St16DOX gene. Plant Physiology and Biochemistry. 2018;**131**:70-77

[68] Thygesen PW, Dry IB, Robinson SP. Polyphenol oxidase in potato (A multigene family that exhibits differential expression patterns). Plant Physiology. 1995;**109**:525-531. DOI: 10.1104/pp.109.2.525

[69] Boeckx T, Webster R, Winters AL, Webb KJ, Gay A, Smith AHK. Polyphenol oxidase-mediated protection against oxidative stress is not associated with enhanced photosynthetic efficiency. Annals of Botany. 2015;**116**:529-540. DOI: 10.1093/aob/mcv081

[70] Werij JS, Kloosterman B, Celis GC, Vos CHRD, America T, Visser RGF, et al. Unravelling enzymatic discoloration in potato through a combined approach of candidate genes, QTL, and expression analysis. Theoretical and Applied Genetics. 2007;**115**:245-252. DOI: 10.1007/S00122-007-0560-Y

[71] Richael CM. Development of the genetically modified Innate® potato. In: Goldman I, editor. Plant Breeding Reviews. John Wiley & Sons, Ltd; 2020. pp. 57-78. DOI: 10.1002/9781119717003.ch3

[72] Ye M, Peng Z, Tang D, Yang Z, Li D, Xu Y, et al. Generation of self-compatible diploid potato by knockout of S-RNase. Nature Plants. 2018;**4**(9):651-654. DOI: 10.1038/s41477-018-0218-6

[73] Butler NM, Jansky SH, Jiang J. First-generation genome editing in potato using hairy root transformation. Plant Biotechnology Journal. 2020;**18**(11):2201-2209. DOI: 10.1111/pbi.13376

[74] Chen JS, Dagdas YS, Kleinstiver BP, Welch MM, Sousa AA, et al. Enhanced proofreading governs CRISPR–Cas9 targeting accuracy. Nature. 2017;**550**:407-410. DOI: 10.1038/natur e2426 8

[75] Tang X, Liu G, Zhou J, Ren Q, You Q, Tian L, et al. A large-scale whole genome sequencing analysis reveals highly specific genome editing by both Cas9 and Cpf1 nucleases in rice. Genome Biology. 2018;**19**(1):84

[76] Ricroch A, Clairand P, Harwood W. Use of CRISPR systems in plant genome editing: Toward new opportunities in agriculture. Emerging Topics in Life Sciences. 2017;**1**:169-182. DOI: 10.1042/ETLS201700 85

Chapter 5

Harvesting in Progress: The Crucial Role of Genetically Improved Crops in Latin America

Beatriz Xoconostle-Cázares, Laura Claret Triana Vidal, Yoatzin Guadalupe Domínguez-Fernández, Rosa Obando-González, América Padilla-Viveros and Roberto Ruiz-Medrano

Abstract

Crop genetic improvement in Latin America is necessary to address the region's agricultural challenges and to enhance food security. The use of advanced biotechnological techniques, such as genetic engineering and molecular breeding, should enable the development of crops with improved traits tailored to the unique agroecological conditions prevalent in the region, similar to the observed impact of improved germplasm in leading countries using transgenic or edited plants. Research has focused on enhancing key agricultural traits, including tolerance to abiotic stresses, such as drought and salinity, resistance to pests, and herbicide resistance. However, other modifications designed to cope with emergent diseases and increase in nutritional content key nutrients such as vitamins and proteins should be addressed. Despite the benefits of genetic improvement, challenges, such as public perception, heavy regulatory frameworks, and a deficient communication on the benefits of these technologies, persist. Collaborative efforts among scientists, policymakers, and the public are essential to overcome these challenges. Through the application of innovative biotechnological tools, scientists are crafting crops with enhanced biotic and abiotic resistance, productivity, and nutritional value. As Latin America continues to grapple with the complexities of a changing climate and the imperative to feed a growing population, genetic improvement stands as a crucial ally in the pursuit of a sustainable and resilient agriculture.

Keywords: genetic improvement, biotechnology, *Agrobacterium tumefaciens*, gene editing, integrated crop management

1. Introduction

1.1 Agriculture in one of the most diverse regions on the planet

In 2022, the population in Latin America and the Caribbean reached 659,310,564 [1]. Globally, it is estimated that the population will reach 10 billion by 2050, with a demand for cereals of 1 billion tons and an additional 200 million tons of livestock products per year [2]. It is calculated that in the region, there are 576 million hectares of agricultural land, contributing 14% to the world's agricultural production and accounting for 23% of net agricultural exports. Farmers number over 15 million, and sometimes whole families are involved in agricultural activities. The region has 23.4% of forested areas, including 31% of the planet's freshwater. It is paramount to mention that 50% of the world's biodiversity is present in the region [3]. The region is not immune to limitations in agricultural land; indeed, demographic pressure has caused the conversion of land to housing and has reduced cultivable land area, which is currently estimated at 11% of the Earth's surface and utilizes 70% of freshwater from underground aquifers, rivers, and lakes. In addition to this problem, we must consider that climate change increases vulnerability to food security due to the emergence of new pests and diseases. According to National Aeronautics and Space Administration (NASA), the average surface temperature of the Earth in 2023 was the warmest ever recorded, 1.2 degrees Celsius above NASA's reference period average (from 1951 to 1980) [4]. This extreme heat was associated with forest fires, rising sea levels, and the emergence of pathogens affecting plants, animals, and humans. In the same year, the climatic phenomenon ENSO (also known as El Niño) affected and continues into 2024, although in a neutral ENSO phase with a 79% probability in the second semester [5]. In a conservative risk analysis, considering that the population of Latin America supports its growth rate, food self-sufficiency is a challenge. Projections on global food supply assume that 60% of more production in 2050 will have to result from improved yields in agriculture by 90%, and only 10% from expansion of cultivated areas [6]. Therefore, improvements will have to come from the effective use of water, rational and efficient use of agrochemicals, the use of technologies to predict meteorological phenomena, and predominantly the use of genetically improved genotypes with greater genetic diversity, obtained through available biotechnological tools [7]. The region comprises 29 countries, which have diverse internal organizations, have faced climate challenges, suffer from high input and service costs, soil fertility loss, difficult access to credit, lack of training and technical assistance, and insufficient infrastructure for production, among several others. Political instability and penetration of organized crime are also widespread.

2. Challenges facing agricultural crops in the region

Food production is affected by factors that limit its production and yield; the main problems faced by producers in the region are described below.

2.1 Climate change

A change in climate is considered when it differs from the pattern recorded in a specific area [8, 9]. Erratic changes in temperature, precipitation, and increased carbon dioxide (CO_2) concentration are examples of climate changes that have a negative impact on agriculture. Therefore, plants tolerant to drought, cold, excessive heat, high irradiation, as well as those with the capacity to fix CO_2 more efficiently, are necessary to mitigate climate change. The strategy for their development and acquisition should include all available improvement tools to allow such plants to be grown on a commercial scale. Undoubtedly, modern biotechnology, including precision biotechnology or gene editing, will play a prominent role in the near future [10, 11].

2.2 Poor soils

The region's soil presents problems due to being alkaline or acidic, high salinity, and low organic matter content. To counteract these problems, strategies to increase the availability of nutrients in these soils are necessary. Undoubtedly, the use of modern fertilizers, as well as the use of microbial bioinoculants, will improve the soil profiles that, currently, hinder optimal plant growth in the region [12].

2.3 Lack of modern fertilizers

In addition to the use of macro- and micronutrients, the use of growth regulators, such as auxins and cytokinins, which significantly increase the vegetative and reproductive growth of crops, is a biotechnological resource underutilized in the region. The production of endogenous beneficial microorganisms is necessary to provide plants with nutrients that impact plant health, as well as their safety by having fewer pesticides used for pest and disease control [13]. The production of organic vegetable crops is not only an important market niche for domestic consumption, but also as an opportunity to export products to high-demand countries, with a significant economic return for the producer [14].

2.4 Lack of pesticides and herbicides with lower residuality

Emerging pathogens have overcome the genetic resistance of current germplasms, and despite a long list of commercial agrochemicals that can potentially limit their spread, it is necessary to count with new active ingredients to address multidrug resistance, as well as compounds with lower residuality and toxicity that are thus environmentally friendly [15]. The use of germplasm resistant to insect pests based on *Bacillus thuringiensis* has allowed for a pause in the evolution of multiresistance to chemical pesticides, with the advantage of reduced use of these chemical agents in our food. Additionally, the use of plants resistant to herbicides with low residuality such as glyphosate is a technically desirable example, which despite its advantages, has been banned in countries like Mexico, although scientific evidence proves its safety. However, in addition to the ban on the use of these herbicides, the use of other agrochemicals with high toxicity and residuality is increasing; such is the case of the use of paraquat replacing the feared glyphosate in weed management tasks [16].

2.5 Drought

Drought is defined as a prolonged period of abnormally dry weather that directly influences human activities [17]. The United Nations Convention to Combat Desertification (UNCCD) classifies drylands according to their aridity index as arid, semiarid, and dry sub-humid. These areas, generically referred to as drylands, are characterized by climatic conditions, such as scarce and irregular precipitation, a significant difference between daytime and nighttime temperatures, soils with low organic matter and moisture, as well as high potential evapotranspiration [18]. According to the UNCCD, 12.1% of the Earth's land surface corresponds to arid zones; 17.7% to semiarid zones, and 9.9% to dry sub-humid zones [19]. Over two billion people live in these areas (approximately one in every three inhabitants of the planet), most of them in developing countries. In addition, dry areas host around 50% of livestock and 44% of the world's agricultural land [20]. Latin America has a marked water deficit affecting its agricultural productivity; however, countries like Chile stand out, where 68% of its territory is classified as arid zone, and in the last decade has experienced the phenomenon of mega-drought, reducing up to 40% the expected average rainfall [21, 22]. Mexico is also suffering from several years of drought, and as of January 2024, 81.8% of its municipalities are under diverse degrees of water stress, of which 63% is moderate to exceptional [23]. Despite this, semi-arid regions of the region have greater technification and productivity, but the balance between production and sustainability is fragile [24]. The lack of comprehensive strategies for efficient water uses and low technification has caused not only social, but also environmental and economic problems. Plants with drought tolerance would be an advantage for producers.

2.6 Floods

Floods are defined as events that, due to precipitation, wave action, storm surge, or failure of a hydraulic structure, cause an increase in the level of the free water surface of rivers or the sea, generating invasion or penetration of water in places where it is not usually present, and generally causing damage to the population, agriculture, livestock, and infrastructure [25].

The record of floods has increased globally, with its main cause being considered the growth of human populations in proximity to local ecosystems with effluents, which exposes populations to the risk of flooding during rainy seasons [26]. The loss of flooded crops generally occurs due to root anoxia, as well as invasion by opportunistic microorganisms. In recent years, research has been conducted on the development of new GM crops of agricultural interest that are tolerant to periods of flooding and adverse environmental factors [27].

2.7 Pests

In Mexico, phytosanitary measures for the transport of plants and plant-derived goods are highly regulated to reduce the likelihood of introduction and spread of pests to the country and protect the phytosanitary condition of crops [28]. However, authorities in coordination with academic institutions must be prepared to identify and control the emergence of novel pests, pathogens, and associated diseases in a timely manner. The development of plants resistant to some pests is already a reality, and the benefits of this technology have been proved in various crops and in parts of the world [11]. Economically important crops such as cotton and maize have been genetically transformed by introducing genes encoding proteins with insecticidal action against the main pests of these crops [29]. The main source of insecticidal proteins comes from the bacterium *B. thuringiensis*, with demonstrated activity against pests of lepidoptera, coleoptera, or diptera. Other genes encoding insecticidal proteins such as vegetative insecticidal protein (VIP) from *B. thuringiensis*, lectins, and protease inhibitors have also been evaluated and demonstrated potential for pest control [30]. The application of *Bacillus* spp. and other plant-growth promoting bacteria is a novel strategy that may also be helpful for protecting plants against diverse pathogens.

3. Scenarios for increasing agricultural production

The increase in food production should be the consequence of expanding planting areas, as well as the use of cultural practices such as backyard planting for self-sufficiency, as proposed by various ideological trends. However, it is important to consider that the increase in food demand results not only from a growing population but also from an increase in the demand of specialty foods as well as for non-food products such as biofuels, coupled with the challenges posed by the emergence of novel pests and diseases, climate change, water deficit, decrease of arable land, and soil degradation in cultivated areas. Considering these limitations, there is a need for substantial acceleration in crop improvement and production [31]. Ideally, the increase in food production could come from expanding planting areas; however, arable land and water are two limiting resources not only in Latin America but worldwide. In fact, the conversion of agricultural areas to urban ones in major cities has been observed. Therefore, it is considered that increasing productivity in the field should be the result of a set of science and technology-based strategies that allow for the technification to optimize water, fertilizer, and pesticide use, as well as counting with germplasm displaying resistance or tolerance to biotic and abiotic stress through integrated crop management schemes [32]. Obtaining improved seeds selected from their centers of origin has the advantage of having germplasm with greater genetic diversity, desirable for resisting or mitigating pest or disease attacks, including emerging pathogens. These materials are a genetic source to identify new alleles for pest and disease resistance, as well as other traits for incorporation into cultivated plants and improving their health and safety [33].

3.1 Biotechnology as a tool to solve agricultural problems

Genetically modified plants, including those obtained through new precision technologies such as gene editing, are options that should be considered as part of an integrated strategy to address the agricultural problems of the region and ensure its food security. Considering that biotechnology has had a significant impact over the past 50 years in agriculture and has greatly expanded the objectives of plant breeding programs, there has been a substantial increase in the area of production of biotechnological crops since their first introduction in the 1990s, and it is expected to provide greater benefits with the introduction and commercialization of new germplasms [33]. Biotechnology has shown exceptional performance in weather prediction, early pest, and disease diagnosis, and of course, the generation of genetically improved plants. In the value chain of an agricultural product, it will also be especially important to consider its postharvest, labeling, and traceability of products to reach their consumer markets [34].

3.2 Use of improved plants in Latin America

The use of genetically improved plants is regulated in all countries, and there is no consensus on this issue. Almost all countries are signatories to the Cartagena Protocol and have developed biosafety legal frameworks, which have allowed them to authorize requests for planting genetically modified plants. However, only 58% of countries have an operational regulatory system [35, 36]. In fact, misinformation and commercial interests have fostered polarization in opinions about the use of these biotechnology products. On the other hand, due to the general interest in biotechnology applications, countries with regulatory systems have established a series of technical requirements to authorize their open-field testing in three scenarios: experimental planting, pilot planting, and commercial planting. According to a report from the Food and Agriculture Organization (FAO) Regional Office for Latin America and the Caribbean, in 2021, sustainable projects associated with the use of biotechnology were achieved, although there is substantial lag in the benefits of using biotechnology products [37].

The regulation of edited plants seems to foresee a niche of opportunity, since, if the product is regulated, the plants produced would only contain modifications consisting in one or few base substitutions or deletions, undistinguishable from those obtained through physical or chemical mutagenesis. Furthermore, the latter usually result in multiple changes throughout the genome, as is the case of most agronomically relevant crops obtained since the 1930s to the 1960s. For this reason, gene-edited plants resulting in discrete insertions or deletions would not be subject to the biosafety laws that regulate genetically modified organisms (GMOs). In contrast, if the editing considers insertion of DNA from another organism, it could be subject to the assumptions of the biosafety law. These scenarios have been discussed in various countries in the region, although in some, the issue has yet to be defined, as in the case of Mexico, which has not expressed its position on precision biotechnology products. The future is promising for this technique, which, if not overregulated, will help provide the desired improved plants for the region [36].

In the following section, we describe the production and consumption of GMOs in the countries of Latin America. In Mexico, maize production is a fundamental part of the Mexican diet; it is imperative to improve its productivity since Mexico has become the world's leading importer of maize, increasing food dependency on the United States, from importing 396,000 tons in 1992 to 13.2 million tons in 2022. It is important to mention that Mexico is the main producer in the world of white maize, not so much yellow, which is mainly destined for livestock and industrial sectors. Indeed, the national production of yellow maize is deficient, and if the trend in the production and consumption of animal products, starches, and fructose continues, the shortfall could significantly increase in the short term [38]. Mexico only reports the planting of genetically modified (GM) canola and cotton, although the trend of not using GM plants is concerning, despite having signed trade treaties with Canada and the United States that include the use of genetically improved plants in the Biotechnology sector. However, Mexico has the infrastructure and human resources to produce genetically improved germplasm, including gene editing. Appropriate legislation would make developments evaluated under biosafety conditions to become commercially available for Mexican agriculture. Other technological applications related, but not restricted to agriculture, include the development of projects with efficient and low-emission technology, achieving 1842 agribusinesses to reduce greenhouse gas emissions (GHGs) by 6 million tons of CO_2, in addition to the production of electricity from biomass [39, 40].

An amendment introduced to the Constitution of Ecuador in 2008 has banned the cultivation of transgenic plants. However, a transgenic banana is being developed by researchers at an Ecuadorian university; this germplasm is intended to confer protection against one of the most devastating fungal diseases affecting this crop: *Fusarium oxysporum* f.sp. *cubense* tropical race 4. It will be interesting to see the use of this edited banana worldwide, which will positively impact the bioeconomy of countries where this fruit is considered a basic staple [41, 42]. Additionally, technology has been implemented in Ecuador that allows intelligent climate control directed to 800 farms with 1056 farmers, who have accordingly increased their production. Additionally, they have improved soil quality on 40 thousand hectares, thereby reducing emissions by 20% [39, 42].

Chile is one of the most accepting countries of innovative technologies, with three commercial events featuring Argentine herbicide-resistant canola and soybean, and maize resistant to lepidopteran pests. Chile adopted a regulatory approach for new plant breeding techniques. Parallel technological improvements have resulted in reduced energy usage, optimized pesticide application, and improved water and soil management by producers in the Maule region [43].

In Uruguay, GM maize and soybeans are actively cultivated, and researchers at universities are generating crops with novel traits, nurturing valuable human resources with the potential to produce new varieties for the region. Moreover, they have devised strategies for the effective and efficient use of herbicides in soybeans, resulting in net savings in crop production [39, 44].

Colombia is considered a megadiverse country that has been able to finalize agreements and progress in its public policies to reap the benefits of biotechnological applications, reflected in the commercial authorization of Argentine canola, carnation, cotton, flax, maize, rice, roses, soybeans, sugar beet, and wheat. Colombia has organized technical agroclimatic forums to help producers in predicting climate changes and reducing production costs [36, 40].

Brazil is an impressive example of technological advancements stemming from the Academia. The case of EMBRAPA (Brazilian Agricultural Research Corporation) producing common bean resistant to viruses has garnered global attention for successfully complying with over-regulated GMO legislation [45, 46]. Additionally, it has commercial authorizations for cotton, maize, soybeans, sugarcane, wheat, and eucalyptus. Brazil also stands out for generating new plant varieties obtained through gene editing, positioning it as a regional leader in this field [47].

Argentina is a regional benchmark and promoter of the safe use of genetically modified plants (and animals). The authorized commercial events include alfalfa, cotton, maize, potato, safflower, soybeans, sugarcane, and wheat. Its legislation is the result of consensuses that could serve as example for the development of similar laws in other countries in the region. In 2023, 24 million hectares were planted, representing 12–13% of the global transgenic surface area. Argentina is positioned as the third-largest producer of GM crops worldwide, behind the United States and Brazil. Argentina was the first to market drought-resistant GM wheat [48].

Costa Rica actively participates in regional initiatives on biotechnology and biosafety, and although it produces genetically modified cotton and soybean seeds, they are entirely for export to the country of origin as the seeds are not allowed for local consumption. Currently, about 1600 hectares of cotton and soybeans are planted for the purpose of seed multiplication for export to the United States. Despite this fact, research at public universities aims to obtain gene-edited rice with drought tolerance [49].

Cuba faces important deficits in food security, importing maize and soybeans for animal feed. Although they have produced GM plants, environmental debates have hindered significant plantings. However, they have the potential to produce modified plants, as they have the infrastructure and specialists to generate them [50].

Paraguay has authorized soybeans, maize, and cotton, showing a tendency toward increasing the area of GM crops to occupy significant positions in commercial planting. Paraguay has a planting area of 4 million hectares of GM plants [36].

Bolivia is also a major cultivator of herbicide-resistant soybeans. Legal discussions focus on the coexistence of GM maize with native varieties, considering the benefits in crop yield and production costs [51].

Honduras is the only country in Central America that allows commercial production and field trials of agricultural biotech crops, as it has allowed commercial planting of maize and rice. GM maize is sold for internal consumption and exported to Argentina, Colombia, and United States, while it imports yellow corn and soybean from the United States [52].

Panama has approved GM maize, although no data are available on the productivity of such commercial events [53].

3.3 Genetic improvement techniques

Genetically modified plants are species into which a gene encoding a novel trait has been introduced; this genetic modification aims to increase or decrease the expression of a gene or to introduce a gene encoding a novel trait. If the gene comes from the same plant, it is termed intragenic; if it comes from another plant, cisgenic; and if it comes from another species, transgenic [54, 55].

3.3.1 Process to obtain a genetically modified plant

Genetic modification conducted through genetic engineering differs from conventional plant breeding in precision and specificity. The process of generating a genetically modified crop can be divided into the following stages: 1. Identification and characterization of the gene of interest encoding a new trait. This step involves basic research showing the effectiveness and safety of the trait, such as drought tolerance or pest resistance. 2. Cloning of the gene into a vector with plant regulatory signals, typically using constitutive expression promoters in all plant tissues or differential expression promoters to ensure the trait is present where needed. 3. Genetic transformation and regeneration of the entire plant, involving the introduction of the gene of interest into plant explants through various methods, with the most used being *Agrobacterium*-mediated transformation and bioballistics. 4. Selection of plants encoding the gene of interest, characterizing the number of copies and insertion sites, transcriptional expression levels, and detection of the protein in tissues. 5. Efficacy assays under biosafety conditions and then in open-field trials. 6. Mass multiplication of the variety and/or sexual crosses with compatible elite germplasm for next field use (**Figure 1**) [54, 55].

The bottleneck in genetic transformation is the introduction of the recombinant vector and the regeneration of transformed tissue. There are biological and physical methods for introducing genes from any donor species, thus expanding the possibility of transferring novel traits. Among the most used tools are the natural genetic engineer *Agrobacterium*, particle bombardment, protoplast transformation, and pollen magnetofection, among others [56]. Other transformation techniques have been described; however, the limiting factor is the regeneration of the complete and fertile plant. The biotechnologist selects the transformation tool based on the possibility of regenerating a complete plant from a transformed explant [55, 56].

3.3.2 How to edit a plant genome

In recent years, the generation of improved plants via genome editing, also known as New Breeding Techniques, has accelerated. The methodology involves the use of enzymes that allow cutting the gene to be modified and enabling natural repair mechanisms to result in discrete base insertions or deletions during the repair of the damaged chain. Different methods have been described, as zinc finger nucleases (ZFN), transcription activator-like effector nuclease (TALEN), Meganucleases, although the broadly used method is the CRISPR-Cas9 gene editing system, brilliantly developed

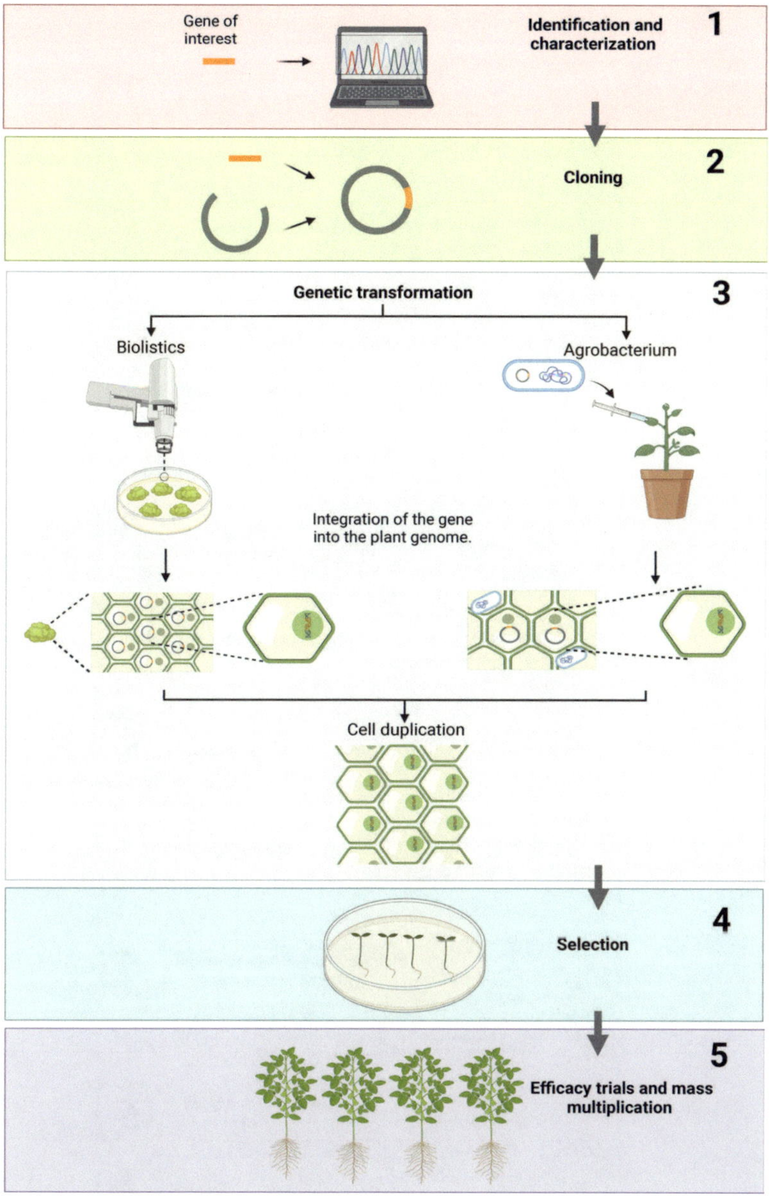

Figure 1.
Process to obtain a genetically modified plant. 1. Identification and characterization of the gene of interest. 2. Cloning of the gene into a plant expression vector. 3. Genetic transformation and regeneration. 4. Selection of GM plants. 5. Efficacy assays and multiplication.

by Doudna and Charpentier [57]. It is also possible to provide a healthy DNA template that serves as a reference for DNA repair and incorporates base changes into the target sequence. It is particularly interesting to note that plants produced with the generated mutations may be free of the editing tool used and be indistinguishable from plants obtained by conventional mutagenesis, a method employed to obtain a large portion of the plants we currently consume worldwide (**Figure 2**) [54].

Among the available genome editing tools, the use of the CRISPR-Cas9 strategy has produced an impressive number of developments, stemming from both

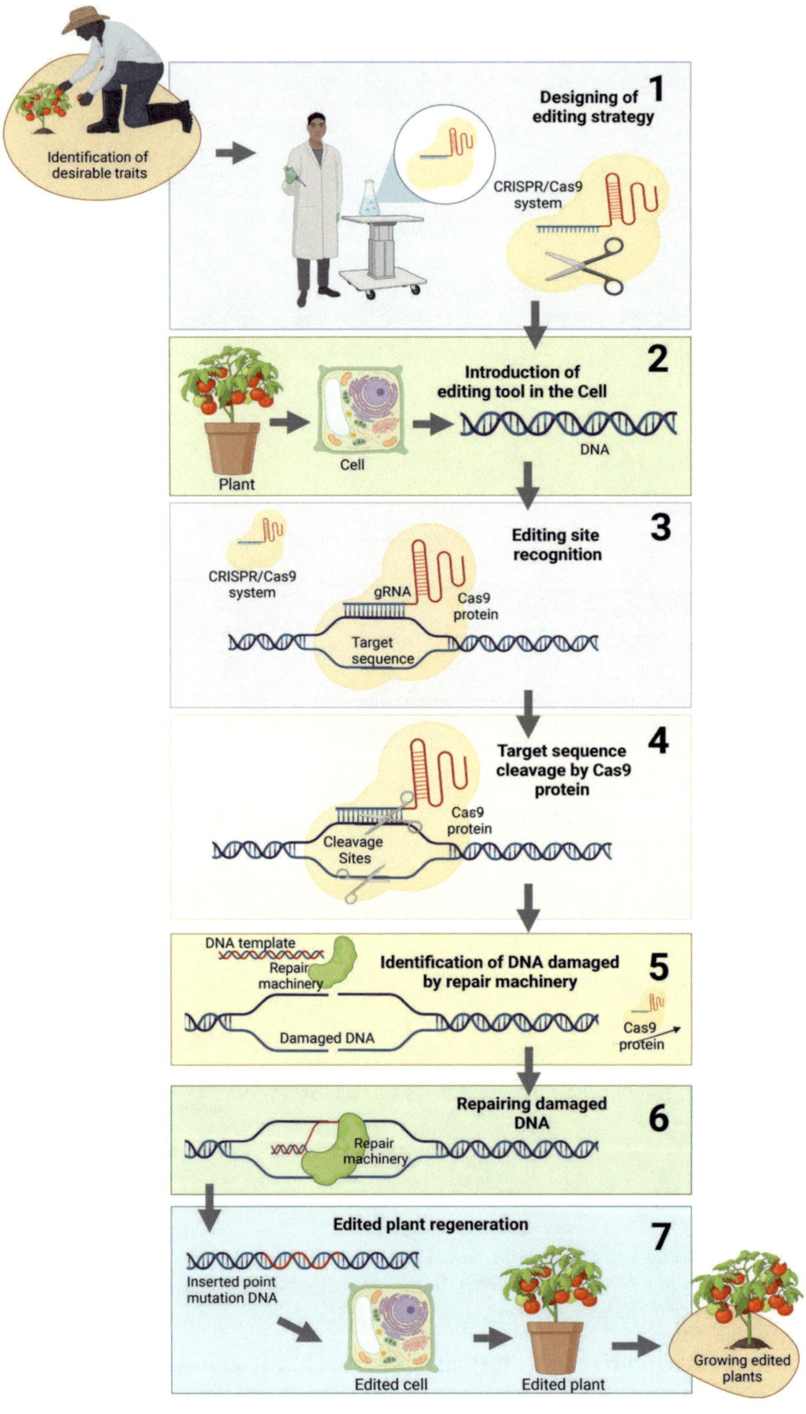

Figure 2.
How to edit a plant gene. 1. Design of a gene editing strategy. 2. Introduction of the editing tool into the cell. 3. Recognition of the editing sequence. 4. Cleavage of target DNA. 5. 6. Repair of DNA by non-homologous end joining (NHEJ) or homologous recombination mechanisms. 7. Edited plant regeneration.

industry and academia. CRISPR-Cas9 stands for Clustered Regularly Interspaced Short Palindromic Repeats and the nuclease Cas9. Originally identified as part of a defense mechanism in some bacteria. The CRISPR-Cas9 system is a newly developed plant breeding method that uses site-specific nucleases to target DNA sequences [57]. This method allows modifications such as base insertion of deletions, provoking gene silencing or editing using DNA templates for repair via homologous recombination (HR), and even allows for transient gene silencing or transcriptional repression. Currently, there is a diversity of editing vectors that allow the design of *ad hoc* complexes, solely with the sequence of the gene of interest. Once the nuclease guided by the complementary RNA has cut the DNA sequence of interest, it must be repaired. If the break is repaired by the non-homologous end-joining (NHEJ) mechanisms, the DNA will likely encode inserted or deleted bases, resulting in permanent silencing of the target gene [54, 58]. Genome editing can be DNA-free if the protein and RNA complex are provided directly into the cells, thus eliminating the need to segregate the editing machinery in the next generation. It is also possible to insert a DNA sequence, also known as "Knock-ins." For this purpose, a donor template encoding the new sequence must be provided, allowing homology-directed repair to incorporate the new sequence. New CRISPR-Cas variants are employed to regulate gene expression, epigenetic modifications, and chromatin interactions [59].

According to the International Service for the Acquisition of Agri-biotech Applications (ISAAA) reports in 2022, the market for plant breeding based on CRISPR in Latin America is projected to increase from 2021 to 2028, with a compound annual growth rate (CAGR) of 12.7% and is expected to reach USD 11.1 million by 2028 [40, 60]. Several international companies are expected to lead this regional development, although numerous public and private universities and research centers are already developing new genotypes, many of which contain insertions or deletions in target genes and are indistinguishable from those obtained by conventional mutagenesis techniques, which consumers are already used to acquiring as part of their diet.

3.4 Commercial developments of GM plants

Currently, the following genetically improved crops are commercially available and have various new characteristics, ranging from those desirable for producers such as resistance to biotic or abiotic stress, herbicides, to those attractive for consumers such as improved oleic profile, flavor, and color (**Table 1**). Alfalfa, apple, Argentine Canola, beans, carnation, chicory, cotton, cowpea, creeping bentgrass, eggplant, eucalyptus, flax, maize, melon, papaya, petunia, pineapple, plum, Polish canola, poplar, potato, rice, rose, safflower, soybean, squash, sugar beet, sugarcane, sweet pepper, tobacco, tomato, and wheat [61, 62].

3.5 Plants obtained with edited genomes

Crops that have been genetically modified using nucleases for different purposes, such as increased resistance to pests, herbicides, improved nutritional profiles, and other desirable traits, are described below.

Soybean (*Glycine max*): Genes have been edited to confer resistance to pests, diseases, and herbicides, as well as to improve nutritional content and yield [63].

Trait	Cultivar	Introduced gene
2,4-D herbicide resistance	Cotton, maize, safflower, soybean	aad-1 or aad-12
Altered lignin production	Alfalfa	ccomt (inverted repeat)
Anti-allergy	Rice	7crp
Coleopteran insect resistance	Maize, potato	Cry34Ab1, cry35Ab1, mcry3A, mcry3A, dvsnf7, cry3Bb1, ipd079Ea, cry3A
Delayed fruit softening	Tomato	Pg (sense, antisense)
Delayed ripening/ senescence	Carnation, melon, pineapple	Sam-k, acc, accd or anti-efe
Dicamba herbicide resistance	Argentine canola, cotton, maize, soybean	dmo
Drought stress tolerance	Maize, soybean, sugarcane, wheat	cspB, Hahb-4, EcBetA, RmBetA
Enhanced photosynthesis/ yield	Maize, soybean	Zmm28, bbx32
Enhanced Provitamin A content	Rice	Crt1, psy1
Fertility restoration	Argentine canola, maize	Bar star or ms45
Foliar late blight resistance	Potato	Rpi-vnt1
Glufosinate herbicide tolerance	Argentine canola, chicory, cotton, maize, Polish canola, rice, safflower, soybean, sugar beet	Bar, pat,
Glyphosate herbicide tolerance	Alfalfa, Argentine & Polish canola, cotton, bentgrass, maize, potato, soybean, sugar beet, sugarcane	cp4 epsps (aroA: CP4), gat4621, goxv247.
Hemipteran insect resistance	Cotton	mCry51Aa2
Imazamox herbicide tolerance	Argentine canola	AtAHAS
Increased ear biomass	Maize	Athb17
Isoxaflutole herbicide tolerance	Cotton, soybean	hppdPF W336
Late blight disease resistance	Potato	RB
Lepidopteran insect resistance	Cotton, cowpea, eggplant, maize, poplar, rice, soybean, sugar cane, tomato,	Cry1F, cry1Ac, cry1F, vip3A(a), cry2Ab2, cry1Ac, cry1Ab-Ac,
Low gossypol	Cotton	dCS
Lowered free asparagine	Potato	asn1
Lowered reducing sugars	Potato	PhL, R1, VInv
Male sterility	Argentine canola, chicory, maize	barnase
Mannose metabolism	Maize, rice	pmi
Mesotrione herbicide tolerance	Soybean	Avhppd-03
Modified alpha-amylase	Maize	Amy797E

Trait	Cultivar	Introduced gene
Modified amino acid	Maize	cordapA
Modified flower color	Carnation, petunia, rose	Dfr, hfl (f3'5'h), bp40 (f3'5'h), dfr-diaca, cytb5, 5AT
Modified fruit color	Pineapple	b-Lyc, e-Lyc, Psy, acc
Modified oil/fatty acid	Argentine canola, safflower, soybean	Te, Lackl-delta12D, Micpu-delta-6D, Pavsa-delta-4D, Pavsa-delta5D, Picpaomega 3D, Pyrco-delta-5E, Pyrco-delta-6E, OtD5E, OtD6D, PirO3D, PlD4D, PpD6E, PsD12D, TcD4D, TcD5D, TpD6E, fad2.2, fatB, gm-fad2–1, D6D
Modified starch/ carbohydrate	Potato	Gbss (AS)
Multiple insect resistance	Cotton, maize, poplar	CpTI, ecry3, IAb, API
Nematode resistance	Soybean	cry14Ab-1.b
Nicotine reduction	Tobacco	NtQPT1 (AS).
Non-browning	Apple	PGAS PPO suppression gene
Oxynil herbicide tolerance	Argentine canola, cotton, tobacco	bxn
Phytase production	Argentine canola	phyA, maize, phyA2, phy02
Reduced black spot	Potato	ppo5
Sulfonylurea herbicide tolerance	Carnation, cotton, flax, maize, soybean	surB, S4-HrA, als, zm-hra, csr1–2, gm-hra,
Tolerance to HPPD inhibiting herbicides	Soybean	hppdPf4Pa
Viral disease resistance	Bean, papaya, plum, potato, squash, sweet pepper, tomato	Ac1 (sense and antisense), prsv_cp, prsv_rep, ppv_cp, pvy_cp, plrv_orf1, plrv orf2, cmv_cp, wmv_cp, zymv_cp.
Volumetric wood increase	*Eucalyptus*	cel1

Modified from Ref. [61].

Table 1.
Expressed traits in genetically modified (GM) cultivars.

Maize (*Zea mays*): Edited genes provide resistance to pests, such as corn borers and corn rootworms, tolerance to herbicides, and enhanced nutritional quality [64].

Cotton (*Gossypium hirsutum*): Edited genes to increase fiber strength, as well as tolerance to herbicides [65, 66].

Canola (*Brassica napus*): Genes have been edited to confer resistance to herbicides and pests, along with modifications to improve oil content and quality [67].

Potato (*Solanum tuberosum*): Edited genes provide resistance to pests and diseases, such as late blight, as well as improved nutritional traits and reduced bruising [68].

Tomato (*Solanum lycopersicum*): Genes have been edited to enhance traits like shelf life, disease resistance, and nutritional content, including increased levels of antioxidants and vitamins [69].

Apple (*Malus domestica*): Edited genes are used to reduce browning after slicing and improve disease resistance [70].

Rice (*Oryza sativa*): Genes have been edited to enhance traits such as pest and disease resistance, tolerance to environmental stresses, and improved nutritional content [71, 72].

Wheat (*Triticum aestivum*): Edited genes provide resistance to pests and diseases, along with improved yield, nutritional quality, and tolerance to environmental stresses [73].

Common mushrooms (*Agaricus bisporus*) with browning prevention (not a plant) are commercially available [74].

4. Benefits of biotechnology products

Since their commercialization in 1996, the benefits related to increased plant productivity, environmental impact, health, and poverty reduction have been documented in countries that have adopted their cultivation. Over the period 1996 to 2020, the estimated economic income increased by $261.3 billion US dollars, considering the main crops (soybean, corn, cotton, and canola). This equates to an average farm income gain across all GM crops grown in this period of about $112/ha. In 2020, the farm income gains were $18.8 billion (average of $103/ha) [39]. The global adoption of GM cotton also displayed a positive impact, as farmers registered a gross farm income gain of about $134.8 million and in the 1997–2020 period [75]. Cost savings in Mexico represented a 29% increase in production compared to conventional varieties in the year 2000 [76]. In general terms, in 2008, the production of the four main crops in the world increased by 29.6 million metric tons (10.1 million tons of soybeans, 17.1 million tons of maize, 0.6 million tons of canola, and 1.8 million tons of cotton) thanks to the cultivation of GM crops [77]. Economic estimates in Argentina report revenues of $19.7 billion since the adoption of herbicide-tolerant soybeans (1996–2006); $482 million for *B. thuringiensis* (Bt) maize and $19.7 million for insect-resistant cotton in the period between 1998 and 2005, giving a total of $20.2 billion in gross revenues for these three crops in the country. Additionally, it is estimated that the adoption of herbicide-tolerant soybeans has contributed to the creation of around one million jobs, being a 36% increase in employment rates in the period 1996–2006 [78]. The first example of drought-tolerant wheat was developed in Argentina, and with the same technology, HB4 transgenic soybeans were already approved for use in the United States, Brazil, Paraguay, Canada, and China in 2022 [79]. Brookes (2022) analyzed for the Antama Foundation the agricultural environmental impact of GM crops in the period 1996–2020, concluding they have beneficially impacted global food, feed, and fiber production by almost one billion tons, with a reduction of its environmental footprint over 17%, calculated in a reduction of carbon emissions by 39,100 million kilograms, saving 14,700 million liters of fuel, similar to removing 25.9 million cars. From 1996 to 2020, the benefit of global net agricultural income was $ 288.54 USD billion, equivalent to an average increase in income of $ 123 USD per hectare. Globally, insect-resistant cotton and maize crops increased yields by an average of 17.7 and 14.5%, respectively, compared to conventional production of equivalent crops. Over 25 years of grown GM crops, global production has increased by 330 million tons of soybeans, 595 million tons of maize, 37 million tons of cotton fiber, 15.8 million tons of rapeseed, and 1.9 million tons of sugar beet. If transgenic crops had not been available to farmers in 2020, additional 11.6 million hectares of soybeans, 8.5 million hectares of maize, 2.8 million hectares of cotton, and 0.5 million hectares of rapeseed would have been needed. These crops have allowed 23.4 million hectares not to be dedicated to agriculture [39, 40].

5. Desirable examples of improved plants in Latin America

Several plant cultivars could benefit from improvement through gene editing or other breeding techniques to address specific challenges faced by farmers in the region. Cultivars that could be targeted for improvement include:

Maize is a staple crop in Latin American countries, improving traits, such as drought tolerance and resistance to may beetles (*Phyllophaga* spp.), could significantly benefit farmers. In addition, a desirable trait is the resistance to infestation by *Aspergillus* spp. in stored seeds, while the high accumulation of aflatoxins in stored maize is currently provoking health problems in consumers.

Soybean cultivation is widespread in Latin America, and enhancing traits, such as herbicide tolerance, reduction of infestation by white fly, and resistance to fungal pests, caused by *Phakopsora pachyrhizi* could lead to increased productivity and profitability for farmers.

Rice is an essential food crop in Latin America, improving its tolerance to flooding or drought could help ensure stable rice production in the region. Golden rice produced through genetic engineering to biosynthesize beta-carotene, a precursor of vitamin A, is a highly desirable development in the region, mainly for children, pregnant women, and elderly population. This crop started to be grown in 2021 in the Philippines, the first country in the world to approve Golden Rice for commercial propagation [80].

Beans are a valuable source of protein and nutrients in Latin American diets. Enhancing productivity could contribute to improving food security and nutrition for local communities. In addition, softer beans would be desirable, as they need to be cooked for longer times to be edible. The hardness of the seeds is discouraging the population from consuming beans, due to the time and energy spent to cook the seeds.

Potatoes are an important staple crop not only in countries like Peru and Bolivia, but are also highly consumed in the region. Improving traits, such as disease resistance, drought tolerance, resistance to nematode infestation, and yield potential, could help safeguard potato production and enhance farmer livelihoods.

Wheat cultivation is expanding in Latin America, improving traits such as yield, disease resistance, and tolerance to heat and drought could support sustainable production in the region. The International Maize and Wheat Improvement Center (CIMMYT) in Mexico is performing an impressive work in providing new genotypes of wheat, maize, and other cultivars adapted to the region [81].

Cotton cultivation is significant in Mexico, Brazil, and Argentina, enhancing traits such as lepidopteran pest resistance, fiber quality, and drought tolerance could support the textile industry, in which the postharvest technology is well established.

Sugarcane is an important crop for sugar production and biofuel, as proven by Brazil for many years, with positive impacts on economy and environment. Improving traits such as yield, disease resistance, and drought tolerance are desirable, as well as the adoption of ethanol and other more energetic alcohols in the region.

Coffee is a vital cash crop in many Latin American countries. Enhancing traits such as yield, disease resistance, coleopteran resistance, and tolerance to climate change could help coffee farmers adapt to changing environmental conditions and market demands. The infestation by nematodes is also a concern in coffee-growing areas.

Bananas are important food crops in Latin America. Improving fungal disease resistance is a key request in the region, currently threatened by *Fusarium oxysporum* f.sp. *cubense* tropical race 4. In addition, postharvest shelf life could help ensure stable production and supply during its value chain.

Pepper is an important crop in Mexico, but also appreciated worldwide nowadays. Insect pest resistance is the most urgent request, including resistance to viral infection, vectored by insects.

Cassava is a staple food around the world, adopted in Latin America, the culinary use of cassava roots and leaves is held back by the presence of cyanogenic glycoside compounds. Improved cassava with lower content of these harmful substances is in demand in the region.

6. Conclusions

Twenty-eight years after the appearance of genetically modified crops, a significant number of them have been released for commercial production with satisfactory results worldwide; proving their quality, safety, and absence of environmental effects. Latin America is a region with asymmetrical adoption of biotechnology products, although with specific demands due to preferences in its diet. Access to biotechnology products also needs to foster national research capacity to generate, evaluate, and adapt innovations according to local needs, support academic groups and public institutions in the country, have clear and appropriate regulations with reliable and transparent biosafety procedures, and have adequate intellectual property rights policies, as well as the integration of academic, industrial, and governmental actors in decision-making regarding the implementation, development, and commercialization of genetically improved crops.

Acknowledgements

We thank the members of our Cinvestav Applied Sciences group for their valuable opinions. Our research is supported by CONAHCYT CF-2023-G-731 to BXC.

Conflict of interest

The authors declare no conflict of interest.

Author details

Beatriz Xoconostle-Cázares[1,2]*, Laura Claret Triana Vidal[1], Yoatzin Guadalupe Domínguez-Fernández[1], Rosa Obando-González[1], América Padilla-Viveros[2] and Roberto Ruiz-Medrano[1,2]

1 Department of Biotechnology and Bioengineering, México

2 Graduate Program in Scientific and Technological Development for Society, Center for Research and Advanced Studies of the National Polytechnical Institute, Av. Instituto Politécnico Nacional, CDMX, México

*Address all correspondence to: bxoconos@cinvestav.mx

References

[1] The World Bank. Population, Total - Latin America & Caribbean [Internet]. Washington, D.C., USA: The World Bank; 2024. Available from: https://data.worldbank.org/indicator/SP.POP.TOTL?locations=ZJ [Accessed: February 20, 2024]

[2] Alexandratos N, Bruinsma J. World Agriculture towards 2030/2050: The 2012 Revision [Internet]. Rome, Italy: World Agriculture; 2012. Available from: www.fao.org/economic/esa [Accessed: February 20, 2024]

[3] FAO. Hacia una agricultura sostenible y resiliente en América Latina y el Caribe-Análisis de siete trayectorias de transformación exitosas. Rome, Italy: FAO; 2021. DOI: 10.4060/cb4415es

[4] Karen Fox, Katherine Rohloff. NASA Analysis Confirms 2023 as Warmest Year on Record. [Internet]. 2024. Available from: https://www.nasa.gov/news-release/nasa-analysis-confirms-2023-as-warmest-year-on-record/ [Accessed: February 27, 2024]

[5] National Oceanic and Atmospheric Administration. El Niño and the Southern Oscillation (ENSO) [Internet]. Washington, D.C., USA: National Oceanic and Atmospheric Administration; 2024. Available from: https://www.ncei.noaa.gov/access/monitoring/enso/technical-discussion [Accessed: February 27, 2024]

[6] FAO. Integrated Crop-Livestock Systems for Development (IC-LSD) Sustainable Production Intensification [Internet]. Rome, Italy: FAO; 2024. Available from: https://www.fao.org/agriculture/crops/thematic-sitemap/theme/ic-lsd/en/ [Accessed: February 28, 2024]

[7] Bruinsma J. The resources outlook: By how much do land, water and crop yields need to increase by 2050? In: Looking Ahead in World Food and Agriculture Perspectives to 2050. FAO; 2011. pp. 270-274. Available from: https://www.fao.org/publications/card/es/c/f70d2098-c9ad-5b05-8972-cf987e74d681/

[8] Oreskes N. The scientific consensus on climate change. Science (1979). 2004;**306**(5702):1686-1686. DOI: 10.1126/science.1103618

[9] Oreskes N. The scientific consensus on climate change: How do we know We're not wrong? In: Lloyd Elisabeth A, Winsberg E, editors. Climate Modelling: Philosophical and Conceptual Issues [Internet]. Cham: Springer International Publishing; 2018. pp. 31-64. DOI: 10.1007/978-3-319-65058-6_2

[10] Ronald P. Plant genetics, sustainable agriculture and global food security. Genetics. 2011;**188**(1):11-20. DOI: 10.1534/genetics.111.128553

[11] Kovak E, Blaustein-Rejto D, Qaim M. Genetically modified crops support climate change mitigation. In: Trends in Plant Science. Vol. 27. Amsterdam, The Netherlands: Elsevier Ltd; 2022. pp. 627-629. DOI: 10.1016/j.tplants.2022.01.004

[12] Riquetti NB, Mello CR, Leandro D, Guzman JA. Climate change projections of soil erosion in South America in the XXI century. Geoderma Regional. 2023;**33**:e00657. DOI: 10.1016/j.geodrs.2023.e00657

[13] Secretaría de Agricultura y Desarrollo Rural. Biofertilizantes: rendimiento y nutrimentos para las plantas y suelos [Internet]. Mexico City, Mexico: Secretaría de Agricultura y

Desarrollo Rural; 2022. Available from: https://www.gob.mx/agricultura/articulos/biofertilizantes-rendimiento-y-nutrimentos-para-las-plantas-y-suelos [Accessed: February 28, 2024]

[14] Instituto Interamericano de Cooperación para la Agricultura. Agricultura orgánica en ALC crece en producción y exportación [Internet]. San José, Costa Rica: Instituto Interamericano de Cooperación para la Agricultura; 2016. Available from: https://iica.int/es/prensa/noticias/agricultura-organica-en-alc-crece-en-produccion-y-exportacion [Accessed: February 28, 2024]

[15] Devilliers SM, Hoisington DA. The trends and future of biotechnology crops for insect pest control. African Journal of Biotechnology [Internet]. 2011;**10**(23):4677-4681. Available from: https://oar.icrisat.org/104/1/dave2011.pdf

[16] Alcántara-de la Cruz R, Cruz-Hipolito HE, Domínguez-Valenzuela JA, De Prado R. Glyphosate ban in Mexico: Potential impacts on agriculture and weed management. Pest Management Science. 2021;**77**(9):3820-3831. DOI: 10.1002/ps.6362

[17] National Oceanic and Atmospheric Administration (NOAA). Drought in America: Slow Moving, Far Reaching [Internet]. Washington, D.C., USA: National Oceanic and Atmospheric Administration (NOAA); 2016. Available from: https://www.noaa.gov/explainers/drought-in-america-slow-moving-far-reaching [Accessed: February 28, 2024]

[18] Davies J. The land in drylands: Thriving in uncertainty through diversity [Internet]. 2017. Available from: chrome-extension://efaidnbmnnnibpcajpcglclefindmkaj/https://www.unccd.int/sites/default/files/2018-06/15.%20The%2BLand%2Bin%2BDrylands__J_Davies.pdf [Accessed: February 28, 2024]

[19] UNNCCD. Convention to Combat Desertification. Washington, D.C., USA: United Nations; 2011. Available from: https://www.unccd.int/events/other/2011-world-day-combat-desertification [Accessed: February 23, 2024]

[20] SEMARNAT. Suelos. In: Informe de la Situación del Medio Ambiente en México Compendio de Estadísticas Ambientales Indicadores Clave y de Desempeño Ambiental [Internet]. 2012th ed. 2013. Available from: https://apps1.semarnat.gob.mx:8443/dgeia/informe_12eng/03_suelos/cap3_1.html [Accessed: February 28, 2024]

[21] Ocampo-Melgar A, Lutz-Ley A, Zúñiga A, Cerda C, Goirán S. Latin American drylands: Challenges and opportunities for sustainable development. Metode. 2023;**13**:33-39. DOI: 10.7203/metode.13.21458

[22] Garreaud RD, Boisier JP, Rondanelli R, Montecinos A, Sepúlveda HH, Veloso-Aguila D. The Central Chile mega drought (2010-2018): A climate dynamics perspective. International Journal of Climatology [Internet]. 2020;**40**(1):421-439. DOI: 10.1002/joc.6219

[23] CONAGUA. CONAGUA [Internet]. 2024. Available from: https://www.gob.mx/conagua/ [Accessed: February 25, 2024]

[24] Zuniga-Teran AA, Mussetta PC, Lutz Ley AN, Díaz-Caravantes RE, Gerlak AK. Analyzing water policy impacts on vulnerability: Cases across the rural-urban continuum in the arid Americas. Environmental Development [Internet]. 2021;**38**:100552. DOI: 10.1016/j.envdev.2020.100552

[25] Centro Nacional De Prevención De Desastres (CENAPRED). Inundaciones

[Internet]. Mexico City, Mexico: CENAPRED; 2021. Available from: www.gob.mx/cenapred [Accessed: February 27, 2024]

[26] Tellman B, Sullivan JA, Kuhn C, Kettner AJ, Doyle CS, Brakenridge GR, et al. Satellite imaging reveals increased proportion of population exposed to floods. Nature. 2021;**596**(7870):80-86. DOI: 10.1038/s41586-021-03695-w

[27] Khalil MI, Hassan MM, Samanta SC, Chowdhury AK, Hassan MZ, Ahmed NU, et al. Unraveling the genetic enigma of rice submergence tolerance: Shedding light on the role of ethylene response factor-encoding gene SUB1A-1. Plant Physiology and Biochemistry. 2024;**206**:108224. DOI: 10.1016/j.plaphy.2023.108224

[28] Diario Oficial de la Federación. Reglamento de la Ley Federal de Sanidad Vegetal [Internet]. 2016. Available from: https://www.dof.gob.mx/nota_detalle.php?codigo=5444943&fecha=15/07/2016#gsc.tab=0 [Accessed: February 27, 2024]

[29] Razzaq A, Zafar MM, Ali A, Li P, Qadir F, Zahra LT, et al. Biotechnology and solutions: Insect-pest-resistance management for improvement and development of Bt cotton (*Gossypium hirsutum* L.). Plants [Internet]. 2023;**12**(23):4071. DOI: 10.3390/plants12234071

[30] Gupta M, Kumar H, Kaur S. Vegetative insecticidal protein (Vip): Potential contender from *Bacillus thuringiensis* for efficient management of various detrimental agricultural pests. Frontiers in Microbiology. 2021;**12**:659736. DOI: 10.3389/fmicb.2021.659736

[31] FAO. The Future of Food and Agriculture and Challenges. Washington, USA: Food and Agriculture Organization of the United Nations; 2017. Available from: https://www.fao.org/3/i6583e/i6583e.pdf

[32] Arús P. Agriculture of the future science and technology for sustainable agricultural development. Metode. 2020;**2020**(10):33-39. DOI: 10.7203/metode.10.12546

[33] Kumari SG, Kumar PL, Moukahel A, El-Miziani I. Phytosanitary management of ICARDA's germplasm seed collections for safe movement and better future use. CABI Reviews. 2023;**18**:1. DOI: 10.1079/cabireviews.2023.0045

[34] Das S, Ray MK, Panday D, Mishra PK. Role of biotechnology in creating sustainable agriculture. PLOS Sustainability and Transformation. 2023;**2**(7):e0000069. DOI: 10.1371/journal.pstr.0000069

[35] Araya-Quesada M, Craig W, Ripandelli D. Biosafety of genetically modified organisms in the Latin American and the Caribbean region: Main needs and opportunities for strategic capacity building. AgBioforum. 2012;**15**:77-88. Available from: https://agbioforum.org/wp-content/uploads/2021/02/AgBioForum-15-1-77.pdf

[36] Zarate S, Cimadori I, Jones MS, Roca MM, Barnhill-Dilling SK. Assessing agricultural gene editing regulation in Latin America: An analysis of how policy windows and policy entrepreneurs shape agricultural gene editing regulatory regimes. Frontiers in Bioengineering and Biotechnology. 2023;**11**:1209308. DOI: 10.3389/fbioe.2023.1209308

[37] FAO. News. 2021. FAO Regional Office for Latin America and the Caribbean. [Internet]. 2021. Available from: https://www.fao.org/americas/en [Accessed: February 25, 2024]

[38] SIAP. 2023 agricultura regenerativa, la vía para un futuro sustentable. [Internet]. 2023. Available from: https://www.gob.mx/siap/articulos/panorama-agroalimentario-2023-la-edicion-estadistica-mas-importante-del-sector-agroalimentario

[39] Brookes G. Farm income and production impacts from the use of genetically modified (GM) crop technology 1996-2020. GM Crops & Food. 2022;**13**(1):171-195. DOI: 10.1080/21645698.2022.2105626

[40] Brookes G. Genetically modified (GM) crop use 1996-2020: Impacts on carbon emissions. GM Crops & Food. 2022;**13**(1):242-261. DOI: 10.1080/21645698.2022.2118495

[41] Dale J, Paul JY, Dugdale B, Harding R. Modifying bananas: From transgenics to organics? Sustainability (Switzerland). 2017;**9**:333. DOI: 10.3390/su9030333

[42] FAO. Court Ruling Confirming the Implementation of the Constitutional Measure as Contained in Article 401 of the Ecuadorian Constitution, a Legal Measure That Protects and Enforces Farmers' Rights to Seeds and Crops That Are GM-Free. Washington, D.C., USA: FAO; 2022. Available from: https://www.fao.org/3/cc0994en/cc0994en.pdf

[43] Sánchez MA. Chile as a key enabler country for global plant breeding, agricultural innovation, and biotechnology. GM Crops & Food. 2020;**11**(3):130-139. DOI: 10.1080/21645698.2020.1761757

[44] GUB.UY. Ministerio de Ganadería, Agricultura y Pesca [Internet]. 2024. Available from: https://www.gub.uy/ministerio-ganaderia-agricultura-pesca/ [Accessed: February 25, 2024]

[45] Ibrahim AB, Aragão FJL. RNAi-mediated resistance to viruses in genetically engineered plants. Plant Gene Silencing. 2015;**1287**:81-92. DOI: 10.1007/978-1-4939-2453-0_5

[46] Daniel Norero. The story behind the 100% public GM bean reaching Brazilian plates. 2021 The story behind the 100% public GM bean reaching Brazilian plates. Available from: https://geneticliteracyproject.org/2021/08/31/the-story-behind-the-100-public-gm-bean-that-reaches-brazilian-plates/ [Accessed: February 25, 2024]

[47] Daniel Norero. Latin America: a biotech laboratory and world champion in GMOs and gene editing. 2023. Available from: https://allianceforscience.org/blog/2023/02/latin-america-a-biotech-laboratory-and-world-champion-in-gmos-and-gene-editing/ [Accessed: February 24, 2024]

[48] Nat Biotechnol. Argentina First to Market with Drought-Resistant gM Wheat. Vol. 39. Nature BIotecHNology News; 2021. p. 652. DOI: 10.1038/s41587-021-00955-y

[49] Hernández-Soto A, Echeverría-Beirute F, Abdelnour-Esquivel A, Valdez-Melara M, Boch J, Gatica-Arias A. Rice Breeding in the new era: Comparison of useful agronomic traits. In: Vol. 27, Current Plant Biology. Amsterdam, The Netherlands: Elsevier B.V.; 2021. DOI: 10.1016/j.cpb.2021.100211

[50] Genetic Literacy Project. Cuba: Crops/Food. Global Gene Editing. Regulation Tracker. Cincinnati, OH, USA: Genetic Literacy Project; 2024. Available from: https://crispr-gene-editing-regs-tracker.geneticliteracyproject.org/cuba-crops-food/ [Accessed: February 23, 2024]

[51] Smale M, Ybarnegaray RP. A Case of Resistance: Herbicide-tolerant Soybeans in Bolivia [Internet]. 2019. Available from: https://www.researchgate.net/publication/232271437

[52] USDA. Agricultural Biotechnology Annual. Foreign Agricultural Service. Washington, D.C., USA: USDA; 2023. Available from: https://fas.usda.gov/ [Accessed: February 22, 2024]

[53] USDA. Agricultural Biotechnology Annual [Internet]. Washington, D.C., USA: PANAM; 2022. Available from: https://fas.usda.gov/data/panama-agricultural-biotechnology-annual-5 [Accessed: February 27, 2024]

[54] Gao C. Genome engineering for crop improvement and future agriculture. In: Cell. Vol. 184. Amsterdam, The Netherlands: Elsevier B.V; 2021. pp. 1621-1635. DOI: 10.1016/j.cell.2021.01.005

[55] Kumar K, Gambhir G, Dass A, Tripathi AK, Singh A, Jha AK, et al. Genetically modified crops: Current status and future prospects. In: Planta. Vol. 251. New York City, USA: Springer; 2020. DOI: 10.1007/s00425-020-03372-8

[56] Sandhya D, Jogam P, Allini VR, Abbagani S, Alok A. The present and potential future methods for delivering CRISPR/Cas9 components in plants. Journal of Genetic Engineering and Biotechnology. 2020;**18**:25. DOI: 10.1186/s43141-020-00036-8

[57] Jinek M, Chylinski K, Fonfara I, Hauer M, Doudna JA, Charpentier E. A programmable dual-RNA-guided DNA endonuclease in adaptive bacterial immunity [internet]. Science. 2012;**337**(6096):816-821. DOI: 10.1126/science.1225829

[58] Mali P, Yang L, Esvelt KM, Aach J, Guell M, Norville JE, et al. RNA-guided human genome engineering via Cas9. Science. 1979;**2013**(339):823-826. DOI: 10.1126/science.1232033

[59] Pickar-Oliver A, Gersbach CA. The next generation of CRISPR–Cas technologies and applications. Nature Reviews Molecular Cell Biology. 2019;**20**:490-507. DOI: 10.1038/s41580-019-0131-5

[60] ISAAA. CRISPR Plant Market Accelerating in Latin America. Ithaca, NY, USA: ISAAA; 2021. Available from: https://www.isaaa.org/kc/cropbiotechupdate/article/default.asp?ID=19955 [Accessed: February 25, 2024]

[61] ISAAA. ISAAA [Internet]. 2024. Available from: https://www.isaaa.org/blog/entry/default.asp?BlogDate=4/5/2023 [Accessed: February 25, 2024]

[62] ISAAA. 50_Biotech_Bites [Internet]. Ithaca, NY, USA: ISAAA; 2016. Available from: https://www.isaaa.org/blog/entry/default.asp?BlogDate=4/5/2023 [Accessed: February 4, 2024]

[63] Yao D, Zhou J, Zhang A, Wang J, Liu Y, Wang L, et al. Advances in CRISPR/Cas9-based research related to soybean [Glycine max (Linn.) Merr] molecular breeding. Frontiers in Plant Science. 2023;**14**:1247707. DOI: 10.3389/fpls.2023.1247707

[64] de Carvalho Teixeira Yassitepe JE et al. Maize transformation: From plant material to the release of genetically modified and edited varieties. Frontiers in Plant Science. 2021;**12**:766702. DOI: 10.3389/fpls.2021.766702

[65] Janga MR, Campbell LAM, Rathore KS. CRISPR/Cas9-mediated targeted mutagenesis in upland cotton

(*Gossypium hirsutum* L.). Plant Molecular Biology. 2017;**94**(4-5):349-360. DOI: 10.1007/s11103-017-0599-3

[66] Zhang X, Cao J, Huang C, Zheng Z, Liu X, Shangguan X, et al. Characterization of cotton ARF factors and the role of GhARF2b in fiber development. BMC Genomics. 2021;**22**(1):202. DOI: 10.1186/s12864-021-07504-6

[67] Ton LB, Neik TX, Batley J. The use of genetic and gene technologies in shaping modern rapeseed cultivars (*Brassica napus L.*). Genes. 2020;**11**:1-22. DOI: 10.3390/genes11101161

[68] Butler NM, Baltes NJ, Voytas DF, Douches DS. Geminivirus-mediated genome editing in potato (*Solanum tuberosum* l.) using sequence-specific nucleases. Frontiers in Plant Science. 2016;7:1045. DOI: 10.3389/fpls.2016.01045

[69] Čermák T, Baltes NJ, Čegan R, Zhang Y, Voytas DF. High-frequency, precise modification of the tomato genome. Genome Biology. 2015;**16**(1):232. DOI: 10.1186/s13059-015-0796-9

[70] Waltz E. Nonbrowning GM apple cleared for market. Nature Biotechnology. 2015;**33**:326-327. DOI: 10.1038/nbt0415-326c

[71] Li T, Liu B, Spalding MH, Weeks DP, Yang B. High-efficiency TALEN-based gene editing produces disease-resistant rice. Nature Biotechnology. 2012;**30**:390-392. DOI: 10.1038/nbt.2199

[72] Shan Q, Zhang Y, Chen K, Zhang K, Gao C. Creation of fragrant rice by targeted knockout of the OsBADH2 gene using TALEN technology. Plant Biotechnology Journal. 2015;**13**(6):791-800. DOI: 10.1111/pbi.12312

[73] Wang Y, Cheng X, Shan Q, Zhang Y, Liu J, Gao C, et al. Simultaneous editing of three homoeoalleles in hexaploid bread wheat confers heritable resistance to powdery mildew. Nature Biotechnology. 2014;**32**(9):947-951. DOI: 10.1038/nbt.2969

[74] Lin X, Sun DW. Research advances in browning of button mushroom (*Agaricus bisporus*): Affecting factors and controlling methods. Trends in Food Science and Technology. 2019;**90**:63-75. DOI: 10.1016/j.tifs.2019.05.007

[75] Brookes G. Genetically modified (GM) crop use in Colombia: Farm level economic and environmental contributions. GM Crops & Food. 2020;**11**(3):140-153. DOI: 10.1080/21645698.2020.1715156

[76] Traxler G, Godoy-Avila S. Transgenic cotton in Mexico. AgBioForum [Internet]. 2004;7:57-62. Available from: http://hdl.handle.net/10355/175 [Accessed: February 23, 2024]

[77] Brookes G, Barfoot P. Global impact of biotech crops: Environmental effects, 1996-2010. GM Crops & Food. 2012;**3**(2):129-137. DOI: 10.4161/gmcr.20061

[78] Burachik M. Regulation of GM crops in Argentina. GM crops & food. 2012;**3**:48-51. DOI: 10.4161/gmcr.18905

[79] Yankelevich A, Megrath C. Agricultural Biotechnology Annual [Internet]. Argentina. Washington, D.C., USA: USDA; 2023. Available from: https://apps.fas.usda.gov/newgainapi/api/Report/DownloadReportByFileName?fileName=Agricultural%20Biotechnology%20Annual_Buenos%20Aires_Argentina_AR2023-0016.pdf [Accessed: February 22, 2024]

[80] IRRI. International Rice Research Institute [Internet]. 2024. Available from: https://www.irri.org/golden-rice [Accessed: February 19, 2024]

[81] CIMMYT. CIMMYT [Internet]. 2024. Available from: https://www.cimmyt.org/ [Accessed: February 19, 2024]

Chapter 6

Genetically Modified Organisms in Urological Cancer

Le Qu, Cheng Chen, Yaoming Li and Xin Pan

Abstract

Genetically modified organisms (GMOs) have become indispensable tools in pre-clinical research for urological cancer. Through genetic engineering techniques, researchers can modify the genetic composition of organisms, enabling the creation of appropriate experimental animal models that provide a deep insight into the mechanisms of tumorigenesis, progression, and potential therapeutic strategies for urological cancer. In this chapter, we provide a comprehensive overview of the current status of research utilizing GMOs in the investigation of prostate cancer, renal cancer, urothelial cancer, and other urological cancers. Topics covered the development of different genetically modified animal models, and the application of these models in urological cancer research. In addition, the limitations of GMOs in cancer research will be discussed.

Keywords: genetically modified organisms, renal cancer, prostate cancer, urothelial cancer, transgenic animal models

1. Introduction

Urological malignancies, including cancers of the prostate, bladder, kidney, and other organs of the urinary system, pose a rapidly increasing global cancer burden [1]. In 2020, Global Cancer Observatory (GLOBOCAN) data indicated that over 2.5 million individuals were diagnosed with malignant tumors of the urinary system, resulting in nearly 800,000 deaths [2]. Prostate cancer accounts for 29% of all new cancer diagnoses among men, as per the American Cancer Society, with bladder, kidney, and renal pelvis cancers also ranking in the top 10 for incidence. Among women, renal cancer is the 9th most common cancer [3]. Despite advances in diagnostics and treatments, the outlook for patients with advanced urological cancers remains bleak [4–9], underscoring the critical need for more research into their pathogenesis and treatment options.

The use of genetically modified organisms (GMOs), through transgenic technology, has revolutionized biology and medical research [10, 11]. Genetically modified animal models, in particular, have become indispensable for studying intricate biological processes, including development, aging, and disease progression [12–15]. These models stand out from other animal models by more accurately mimicking the genetic and pathological aspects of human urological cancers. Notably, some of these models can spontaneously develop metastatic disease, facilitating study of the micro-environment at various stages of tumor metastasis. They also allow for the assessment of therapeutic effects and the validation of drug targets across different stages of the disease [16–18].

This chapter provides a comprehensive review of the utility of GMOs, particularly genetically modified animal models, in uncovering the pathogenesis of urological cancers and identifying new therapeutic strategies or drug targets.

2. GMOs as essential models in urological cancer research

2.1 Prostate cancer (PCa)

2.1.1 TRAMP model

The transgenic adenocarcinoma of the mouse prostate (TRAMP) model, first reported by N. M. Greenberg in 1995 [19], employs transgenic technology to introduce the androgen-dependent, prostate tissue-specific protein rat probasin (rPB) into fertilized mouse eggs. This process inactivates tumor suppressors p53, retinoblastoma 1 (Rb1), and protein phosphatase 2A (PP2A) in the prostate, leading to PCa development in mice [20, 21]. Characterized by its clinicopathological features and widespread metastasis, the TRAMP model is considered the most detailed PCa model available. By 8 weeks, TRAMP mice show abundant Simian Virus 40 (SV40) large T-antigen markers in the dorsolateral lobes of the prostate, progressing from low-grade to high-grade prostatic intraepithelial neoplasia by 10 weeks. At 12 weeks, distant metastasis begins, with all mice developing metastatic PCa by 28 weeks—a 100% tumor formation rate [22, 23]. Metastases occur in various locations, predominantly in the dorsolateral lobe of the prostate [24]. Due to its extensive metastatic spread [25, 26], the TRAMP model is pivotal for studying PCa's pathogenesis and metastatic behavior [27].

2.1.2 LADY model

The LPB promoter driving the large T-antigen (LADY) model, using the large PB promoter (LPB) to express SV40 Large T-antigen (Tag) without the small t-antigen, contrasts with the TRAMP model to highlight neuroendocrine differences in metastatic lesions [28, 29]. It demonstrates progression from prostatic intraepithelial neoplasia (PIN) with neuroendocrine features to invasive neuroendocrine carcinoma [30]. Metastatic lesions appear at one site by 6 months and at multiple sites by 9–10 months, yet only 14% develop bone metastases. The model typically shows metastasis to lymph nodes, liver, lung, spleen, kidney, and occasionally bones [29, 30] but exhibits limited PCa cell proliferation and less invasiveness compared to TRAMP. It serves as an effective model for studying early PIN changes and tumor stroma [31, 32], making it suitable for exploring the mechanisms of neuroendocrine PCa (NEPC) [28].

2.1.3 Other SV40 T-antigen

Beyond the commonly used TRAMP and LADY models, other transgenic models have been developed using SV40 large T-antigen and small t-antigen, combined with various promoters. Jeffrey E. Green et al. introduced a C3(1) initiator to express SV40 large T-antigen, creating the first PCa genetically modified animal model [33, 34]. This model causes female mice to develop breast cancer at 6 months and male mice to develop prostate hyperplasia by 8 months, which can progress to PIN and PCa, occasionally metastasizing to the lung and bone [35]. This genetically modified animal model facilitates the study of hormone response components *in vivo*, as well as the multistage progression from normal tissue to PCa [36, 37].

Jeffrey I. Gordon et al. developed a model for metastatic PCa using the Cryptidin-2 gene promoter to direct expression of SV40 large T-antigen in prostate neuroendocrine cells, leading to neuroendocrine PCa [38]. The model mice developed PIN at 8 weeks, followed by invasive cancer within 2–4 weeks and metastasis to lymph nodes, liver, lung, brain, and bone within 16 weeks [39]. The tumors were androgen-independent, reca pitulating multiple histopathological features of human PCa while avoiding androgen dependence on neuroendocrine cancer evolution [40, 41]. Compared to the TRAMP model, this model has a shorter tumor induction period and provides new biomarkers for clinical diagnosis of PCa neuroendocrine differentiation.

Jim W. Xuan et al. utilized the knock-in technique, using the prostate secretory protein 94 amino acid (PSP94) gene promoter/enhancer to target and regulate mouse prostate-specific expression of SV40 large T/small t-antigen, establishing a new PCa model, prostate secretory protein-knock-in mouse adenocarcinoma prostate (PSP-KIMAP) [42, 43]. This model recapitulates aspects of human PCa, with PIN observed around 10 weeks, progressing to invasive adenocarcinoma by 24 weeks and accompanied by distant metastases to lymph nodes, lungs, and liver [40]. Compared to TRAMP and LADY models, PSP-KIMAP more accurately simulates human PCa development, exhibiting more stable phenotypes and precise prostate tissue targeting [30, 44]. While the PSP-KIMAP study is not comprehensive, it is worth acknowledging for replicating PCa and serving as a valuable addition to other models, holding broad application prospects.

2.1.4 Pten

Phosphatase and tensin homolog deleted on chromosome ten (Pten) gene, a significant tumor suppressor, is often deficient in various human cancers [45]. In human PCa, Pten deletion is found in 23% of high-grade PIN (HGPIN), 69% of localized cases [46], and 86% of metastatic castration-resistant PCa (CRPC) [47]. This deficiency underscores Pten's crucial role in PCa initiation, leading to the development of Pten-deficient mouse models. Shun you Wang et al. created a model with prostate-specific Pten deletion, revealing that heterozygous mice developed PIN at 10 months, while homozygous deletion resulted in invasive adenocarcinoma by 9 weeks, with metastasis by 12 weeks [48]. However, tumors also appeared in non-prostate tissues [49–51], which could limit the model's specificity for PCa research.

Researchers have explored the role of interleukin-17 (IL-17) in PCa using these models, shedding light on the tumor microenvironment [52, 53]. They also investigated the interaction between tumor suppressors Rb and Pten using double mutant mice with cyclin-dependent kinase (Cdk) inhibitor p18Ink4c and Pten knockout [54]. Results show that double mutant mice have faster and more extensive tumor growth in the prostate anterior and dorsolateral lobes [54]. Additionally, loss of Nkx3.1, combined with Pten deficiency, significantly increases HGPIN incidence, mirroring early stage human PCa [55, 56]. Similarly, DePinho et al. found that prostate-specific deletion of Pten (Pten $^{pc-/-}$) and prostate-specific deletion of Smad4 (Smad4 $^{pc-/-}$)

in mouse models exhibited highly invasive characteristics with deep lymphatic and pulmonary metastases [57]. Based on these findings, the Pten $^{pc-/-}$/Smad4 $^{pc-/-}$ mouse model was constructed and applied to research on combinations of hypoxia-activated prodrug TH-302 and checkpoint blockade [58]. They found that the combination of TH-302 and checkpoint blockade significantly increased the survival of Pten $^{pc-/-}$/ Smad4 $^{pc-/-}$ mice [58]. Moreover, Xin Lu and colleagues also utilized this model to reveal potential pathways for improving immunotherapy in advanced prostate cancer through Pygo2 (Pygopus2 (Pygopus family plant homeo domain (PHD) finger 2))-targeted treatment [59]. Additionally, a model with Pten deletion and speckle-type pox virus and zinc finger protein (Spop) mutation demonstrated the role of Spop mutations in activating phosphatidylinositol-3-kinase/mammalian target of rapamycin (PI3K/mTOR) and androgen receptor (AR) signaling, advancing the understanding of PCa progression [60]. The Pten/Kras (Kirsten rat sarcoma viral oncogene homolog) model reported by David J. Mulholland et al. was found to accelerate prostate cancer progression due to Pten deficiency, with concomitant epithelial-mesenchymal transition (EMT) and extensive metastasis [61].

2.1.5 Myc

Previous studies have suggested that upregulation of Myc may be a critical driving event in the onset and progression of human PCa [62, 63]. Charles L. Sawyers and his team created transgenic mice expressing human c-Myc in the prostate [64], categorized into Hi-Myc and Lo-Myc groups based on androgen sensitivity, with Hi-Myc being androgen-sensitive. The Hi-Myc mice showed accelerated PIN progression compared to the Lo-Myc group [64]. Despite its advantages, the model's lack of transferability is a significant limitation [64]. Interestingly, concurrent Myc overexpression and Pten deficiency in mice resulted in aggressive adenocarcinomas with distant metastases. The research on this animal model has confirmed that Homeobox protein Hox-B13 (Hoxb13) played a pivotal role in the causation of prostate cancer [65, 66]. Combining Hi-Myc mice with PB-Hepsin mice reduced adenocarcinoma progression from 24 to 12 weeks [67]. However, studies show that mTOR inhibitors are ineffective against Myc overexpression mice, suggesting that they may be contraindicated in Myc overexpression PCa patients [68]. Given Myc's early amplification in PCa [62, 69], the Myc transgenic model serves as a foundational tool for studying genetic alterations in PCa progression, often used in conjunction with other PCa-related genes.

N-Myc and L-Myc, members of the Myc family, play roles in PCa development and progression. Mycl amplification is observed in precancerous lesions and early tumors [70], whereas N-Myc is associated with aggressive castration-resistant PCa (CRPC) and NEPC [71]. Etienne Dardenne et al. established a transgenic mouse model overexpressing N-Myc and identified it as an oncogenic driver of NEPC [72]. This model suggests that AR signaling is abolished and Polycomb Repressive Complex 2 (Ezh2) signaling is induced [72]. The model helps identify and validate potential therapeutic targets for treating NEPC, such as Ezh2 and Aurora A kinase.

2.1.6 Cdcp1

Abdullah Alajati et al. developed a mouse model of prostate cancer (PCa) that overexpresses CUB domain-containing protein 1 (Cdcp1, 73]. In this model, 50% of mice exhibited prostate hyperplasia within 4–6 months, which progressed to PIN by 7–9 months and to HGPIN by 14 months [73]. The simultaneous overexpression of

Cdcp1 and knockdown of Pten markedly accelerated the development of metastatic PCa in these mice. Utilizing this model, the researchers identified promising drug treatment strategies for combating metastatic PCa [73].

2.2 Renal cancer

Research highlights the critical tumor-suppressing role of von Hippel-Lindau (Vhl) in clear cell renal cell carcinoma (ccRCC) development [74], notably at tumorigenesis's initial stages [75]. While modern sequencing technologies corroborate this [76], studies indicate that Vhl loss alone does not suffice to induce renal cell carcinoma (RCC) [77–81]. This insight has led researchers to combine Vhl deletion with other oncogenic modifications, aiming to develop an optimal renal cancer model. Following Vhl, Polybromo 1 (Pbrm1) emerges as the second key suppressor gene in renal cancer, with about 40% of ccRCC cases involving Pbrm1 mutations [82–84]. Interestingly, single knockdown of Pbrm1 by some researchers through knockout technology did not result in the expected renal cancer model. However, discoveries of ccRCC with combined mutations in Vhl and Pbrm1 [83, 85], and metastatic ccRCC with mutations in both Brca1-associated protein 1 (Bap1) and Pbrm1 [85, 86], suggest ccRCC may require simultaneous knockdown of two or more genes.

2.2.1 Vhl/Pbrm1

Amrita M Nargund and colleagues have developed a ccRCC mouse model by knocking out both Vhl and Pbrm1 genes via the Ksp-Cre method [87]. This model not only led to renal cyst formation but also successfully induced ccRCC. Mice deficient in both Vhl and Pbrm1 showed a 30% incidence of preneoplastic polycystic kidney diseases by 6–9 months and a 50% ccRCC incidence by 10 months [87]. Similarly, Yi Feng Gu's team created a double knockout model using Paired Box 8 (Pax8)-Cre, resulting in about 85% of mice developing extensive tumors by 9 months, escalating to 100% by 13 months [88], with tumors eventually nearly replacing the kidneys by 16 months [88]. The Vhl/Pbrm1 model serves as a valuable tool for exploring the molecular dynamics of Vhl and Pbrm1 mutations and for conducting drug efficacy tests.

2.2.2 Vhl/Bap1

Subsequent research identified the Bap1 gene as significantly associated with ccRCC, ranking it as the third most important ccRCC-associated gene with a mutation rate of approximately 15% [83, 85, 89]. Bap1 mutations are linked to higher-grade ccRCC, contrasting with the lower-grade associations of Pbrm1 mutations. Shan Shan Wang et al. initially attempted a simultaneous Vhl/Bap1 deletion using Six2-Cre, but mice with Bap1 deficiency died within a month of birth [89]. Later, Yi Feng Gu et al. successfully created a Vhl/Bap1 deletion model with Pax8-Cre, which, alongside the Vhl-Pbrm1 model, elucidates the role of these genes in ccRCC formation and tumor grading. Bap1-deficient tumors tend to be of higher-grade compared to Pbrm1-deficient tumors [88, 90].

2.2.3 Vhl/Trp53/Rb1

Sabine Harlander and colleagues developed a new ccRCC mouse model by deleting Vhl, transformation-related protein 53 (Trp53), and Rb1 in renal epithelial cells [91]. This model provides a basis for studying hypoxia-inducible factor-alpha (HIF-α)

inhibition as a potential ccRCC treatment and offers a platform for drug screening and testing new therapies [91]. However, such triple gene inactivation is rare in human ccRCC, and the model does not exhibit metastases to lung, liver, bone, or brain, highlighting its limitations.

2.2.4 Myc/Vhl/Cdkn2a

Sean T. Bailey and his team created a metastatic renal cancer model by overexpressing Myc and deleting Vhl and cyclin-dependent kinase inhibitor 2A (Cdkn2a), mimicking human ccRCC [92]. Remarkably, about one-third of the mice developed liver metastases [92]. However, the combination of Vhl and Cdkn2a loss, along with Myc activation, is relatively rare in humans, limiting the model's representativeness [92].

2.2.5 Flcn

The folliculin (Flcn) gene, identified from a Birt-Hogg-Dube syndrome (BHD) patient [93, 94], plays a role in causing BHD syndrome, which includes RCC among other diseases. Initially established whole-body Flcn knockout homozygote mice died at the embryonic stage, whereas heterozygous mice had a very late onset of disease with a low and erratic morbidity rate [95–97]. Using Ksp-Cre, researchers established an Flcn-deficient model that successfully mimics RCC [98–100], showing 100% renal cyst incidence and about 70% tumor incidence within 6–7 months [100]. Thus, the Flcn-deficient model can be utilized for molecular mechanism studies and drug testing and screening in renal cancer. Unfortunately, there is a diversity of histologic subtypes of RCC observed in this model, and it is not yet possible to determine whether there are other factors that determine the histologic type of RCC in this model, a question that remains to be investigated.

2.2.6 Other renal cancer models

Zachary S. Morris et al. developed a genetic mouse model with neurofibromatosis type 2 (Nf2) knockout using Villin-Cre [101], observing renal cell hyperplasia and cysts at 15 days, and small renal tumors by 3 months that progressed to invasive renal cancer by 6 to 10 months. This model also exhibited elevated epidermal growth factor receptor (Egfr) expression, indicating that Nf2 inactivation activates the Egfr signaling pathway, promoting renal cancer development [101]. Although Nf2 is not primarily associated with renal cancer, this model presents an early onset opportunity for studying Nf2-related renal cancer pathogenesis and drug screening.

Lorraine J. Gudas et al. reported a mouse model for Vhl renal cancer known as the TRAnsgenic model of Cancer of the Kidney (TRACK). This model is characterized by the specific expression of mutant hypoxia-inducible factor 1-alpha (Hif1a) in the renal proximal tubule cells, closely mirroring many early stage features of ccRCC [102].

Alessia Calcagnì et al. created the first genetic animal model of renal cancer with transcription factor EB (Tfeb) overexpression using cadherin 16 (CDH16)-Cre [103]. This model sheds light on the mechanism of TFE-fusion RCC and suggests a therapeutic approach targeting the Wingless/integrated (WNT) pathway [103]. However, the model mice developed severe renal impairment, an outcome not commonly seen in humans, which may limit the model's utility for therapeutic exploration.

Qiang Hua Hu et al. produced a Wilms tumor (WT) mouse model by increasing Igf2 expression and eliminating WT1, making it the only WT transgenic mouse model closely resembling human tumors [104]. Visible tumors appeared at 9 weeks, providing a valuable tool for testing therapeutic strategies [105].

2.3 Urothelial cancer (UC)

The uroepithelium, one of the body's slowest-renewing epithelia, undergoes unique biological transformations. Oncogene inactivation or activation plays a crucial role in bladder tumorigenesis [106]. With gradual clarification of UC-related molecular mechanisms and maturing molecular biology techniques, GMOs have been successfully established, better reproducing human UC biological behavior at the molecular level. GMOs are widely used to study specific gene functions of Hras (Harvey rat sarcoma viral oncogene homolog), P53, Pten, Rb, fibroblast growth factor receptor 3 (Fgfr3), and epidermal growth factor receptor (Egfr) in bladder cancer development [107].

2.3.1 SV40 T-antigen

Zhong Ting Zhang et al. developed a transgenic mouse model in 1999 using the uroplakin II (UpII) promoter to express SV40 T-antigen urothelially [108]. This model showed that both low and high copy number SV40T transgenic mice develop bladder carcinoma *in situ* (CIS), with high copy number mice progressing to invasive and metastatic transitional cell carcinoma (TCC) [108]. Francisco Ayala de la Peña et al. later created a model using UPK II-Cre that did not progress to invasive tumors [109].

2.3.2 Hras

Hras, the first oncogene identified in human UC [110], led Zhong Ting Zhang et al. in 2001 to establish a mouse model activating Hras in the uroepithelium. This resulted in urothelial hyperplasia and superficial papillary non-invasive bladder tumors [111]. They found that low copy number Hras transgenic mice developed non-invasive lesions, whereas high copy number mice died by 5 months, highlighting the RAS pathway's role in developing low-grade, non-invasive papillary UC [111].

2.3.3 P53

P53, crucial for uroepithelial cell growth control [112], is frequently mutated or deleted in human UC [113]. Jing Gao and his team have demonstrated that the complete loss of p53 is a prerequisite for the activation of Hras to promote the generation of UC via a model constructed by hybridizing active Hras transgenic mice with p53-deficient mice [114]. In addition, Anna M. Puzio-Kuter et al. have shown that combined deletion of p53 and Pten in bladder epithelial cells leads to invasive cancer in a novel mouse model that provides a validating tool to study mTOR inhibitors for the treatment of invasive UC [115].

2.3.4 Pten

Chao Nan Qian and colleagues created a model by specifically knocking out Pten in kidney epithelial cells using Ksp-Cre [116]. About 57% of these mice developed UC of the renal pelvis by 12 months, with significantly increased phosphorylated mTOR levels, suggesting mTOR inhibitors as effective treatments [116].

2.4 GMOs of other urological cancers

Testicular germ cell tumors (TGCTs), the predominant form of testicular cancers among young men, originate from germ cells [117]. James A. Gill et al. developed TGCT in zebrafish by expressing SV40 large T-antigen under the pufferfish lymphocyte-specific protein tyrosine kinase (Flck) promoter, observing TGCT development after a latency of up to one year. Additionally, overexpression of the stem cell leukemia (Scl) gene in zebrafish testis also led to TGCT [118]. These findings demonstrate the viability of studying TGCT *in vivo*. Earlier mouse models were limited to benign tumors, not reflecting the malignant TGCT more commonly seen in young men [119–121]. Amy M. Lyndaker et al. constructed a mouse model of malignant TGCT by localizing Kras activation and Pten inactivation in mouse premeiotic germ cells [122, 123].

Penile cancer, a rare urinary system malignancy, is often preceded by penile intraepithelial neoplasia (PeIN), a precancerous lesion linked to human papillomavirus (HPV) [124, 125]. Beatriz Medeiros Fonseca and colleagues introduced the first mouse model mimicking HPV-related penile cancer [126]. This model was created by treating 10-week-old HPV16 transgenic mice with dimethylbenz(o) anthracene (DMBA) over a 16-week period, leading to the development of HPV-associated penile cancer traits similar to those observed in humans, such as condylomas and PeIN [126]. Their pioneering research has paved the way for further studies into the underlying mechanisms and potential treatment options for penile cancer.

3. Challenges and limitations of GMOs

Advances in genome editing, such as clustered regularly interspaced short palindromic repeats/CRISPR-associated protein 9 (CRISPR/Cas9) [127], face challenges like off-target effects and exogenous mutations, complicating the maintenance of transgenic models and impacting research. The complexity of gene regulation in urological cancer adds to the difficulty of creating accurate models. These challenges hinder the broader application of transgenic models in urological cancer research. Ethical and safety considerations of GMOs also demand careful attention.

4. Potential development trends of GMOs

Improvements in CRISPR/Cas9 and the advent of new editing techniques like the National Institute for Cancer Epidemiology and Research (NICER) and AsCas12f promise enhanced transgenic model efficiency and accuracy [128, 129]. Future urologic oncology research may benefit from interdisciplinary collaboration and integrating technologies like three-dimensional (3D) printing and artificial intelligence (AI), potentially revolutionizing our understanding of urological tumors.

Cancer type	Model	Time of penetrance, Pathology	Characteristics	References
Prostate cancer	TRAMP	10 weeks: LGPIN to HGPIN 12 weeks: distant metastasis begins 28 weeks: 100% metastatic tumor formation	First autochthonous mouse model of PCa	[19]
	LADY	6 months: metastatic tumor at one site 9–10 months: metastatic tumor at multiple sites (14% bone)	Less invasiveness	[28]
	C3(1)	8 months: prostate hyperplasia even PIN and PCa	Hormone response components *in vivo*	[33]
	Cryptidin-2	8 weeks: PIN 10–12 weeks: invasive cancer Within 16 weeks: metastasis to lymph nodes, liver, lung, brain, and bone	Shorter tumor induction period	[38]
	PSP-KIMAP	10 weeks: PIN 24 weeks: invasive AD accompanied by distant metastases to lymph nodes, lungs, and liver	More accurately simulates human PCa	[42]
	Pten	Pten$^{+/-}$ 10 months: PIN Pten$^{-/-}$ 9 weeks: invasive AD 12 weeks: metastasis	Tumors also appeared in non-prostate tissues	[48]
	Pten/p18Ink4c	9 months: Pten$^{+/-}$ /p18$^{-/-}$ HGPIN 12 months: Pten$^{+/-}$ /p18$^{+/-}$ HGPIN or carcinoma	Faster and more extensive PCa growth	[54]
	Pten/Nkx3.1	26–52 weeks: 60% HGPIN 52 weeks: 100% HGPIN >52 weeks: 84% AD, 25% lymph node metastasis	Mirror early stage human PCa	[55, 130]
	Pten/Smad4	7 weeks: LGPIN 11 weeks: invasive PCa 15 weeks: highly aggressive PCa 32 weeks: 100% lymph node, 12% lung metastasis	Highly invasive	[57]
	Ptne/Kras	10 weeks: PIN 20 weeks: AD 40 weeks: Death	EMT and stem-like features	[61]
	Hi-Myc	>13 weeks: PIN >26 weeks: AD	Accelerated PIN progression	[64]
	Hi-Myc/PB-Hepsin	4.5 months: invasive AD	Higher-grade AD	[67]
	N-Myc/Pten	NEPC	Abolish AR signaling	[72]
	Cdcp1	4–6 months: prostate hyperplasia 7–9 months: PIN 14 months: HGPIN	SRC/MAPK pathway activation	[73]

Cancer type	Model	Time of penetrance, Pathology	Characteristics	References
Renal cancer	Vhl/Pbrm1	Ksp-Cre 6–9 months: 30% preneoplastic polycystic kidney 10 months: 50% ccRCC Pax8-Cre 9 months: 85% ccRCC 13 months: 100% ccRCC 13 months: tumor nearly replace kidney	Valuable tool for ccRCC research	[87]
	Vhl/Bap1	Six2-Cre: die within a month of birth Pax-8cre: different grades of ccRCC	Higher-grade tumors	[88]
	Vhl/Trp53/Rb1	10 of 25 develop a total of 64 tumors	No metastasis	[91]
	Myc/Vhl/Cdkn2a	one-third of the mice developed liver metastases	Rare in human ccRCC	[92]
	Flcn	6–7 months: 100% renal cyst, 70% tumor	Diversity of histologic subtypes	[100]
	Nf2	15 days: renal cell hyperplasia and cysts 6–10 months: invasive renal cancer	Nf2-related renal cancer	[101]
	TRACK	ccRCC	Hif1a activation	[102]
	Tfeb	TFE-fusion RCC	First genetic animal models of RCC	[103]
	Igf2	WT	The only WT transgenic mouse model	[104]
Urothelial cancer	SV40 T-Antigen	low copy number: CIS high copy number: metastatic TCC	Develop bladder CIS and TCC	[108]
	Hras	low copy number: urothelial hyperplasia, superficial papillary non-invasive bladder tumors	RAS pathway activation	[111]
	P53/Pten	Invasive cancer	NA	[115]
	Pten	12 months: 57% UC	Develop UC of the renal pelvis	[116]
Testicular germ cell tumors	SV40 T-antigen	>1 year: TGCT	Develop TGCT in zebrafish	[118]
	Scl	TGCT	NA	[118]
	Kras/Pten	TGCT	NA	[122]
Penile cancer	HPV16	16 weeks: condylomas and PeIN	First mouse model mimicking HPV-related penile cancer	[126]

Abbreviations: PIN prostatic intraepithelial neoplasia; PCa prostate cancer; NEPC: neuroendocrine prostate cancer; LGPIN: low-grade prostatic intraepithelial neoplasia; HGPIN: high-grade prostatic intraepithelial neoplasia; AD: adenocarcinoma; AR: androgen receptor; ccRCC: clear cell renal cell carcinoma; WT: Wilms tumor; CIS: carcinoma in situ; TCC: transitional cell carcinoma; UC: urothelial cancer; TGCT: testicular germ cell tumor; NA: not applicable; PeIN: penile intraepithelial neoplasia.

Table 1.
Commonly GMOs in urological cancers.

5. Conclusions

In conclusion, these GMOs have greatly aided understanding of urological cancers' etiology, progression, and metastasis, and are critical for testing novel drug targets, and assessing treatment responses (**Table 1**). They promote urological cancer research and allow researchers to choose suitable models for deep exploration in their fields.

Acknowledgements

This work was supported by the National Natural Science Foundation of China (No. 82173345 and No. 82373154 to Le Qu).

Conflict of interest

The authors declare no conflict of interest.

Author details

Le Qu[1*], Cheng Chen[2], Yaoming Li[3] and Xin Pan[1]

1 Department of Urology, Jinling Hospital, Affiliated Hospital of Medical School, Nanjing University, Nanjing, China

2 Department of Medical Oncology, Jinling Hospital, Affiliated Hospital of Medical School, Nanjing University, Nanjing, China

3 Department of Urology, Daping Hospital, Army Medical University, Chongqing, China

*Address all correspondence to: septsoul@hotmail.com

References

[1] Chunyang L, Xiaoxi Z, Shi Q, Yonghong G, Yonggang Z. Nanomedicine for urologic cancers: Diagnosis and management. Seminars in Cancer Biology. 2022;**86**(Pt 2):463-475. DOI: 10.1016/j.semcancer.2022.05.014

[2] Hyuna S, Jacques F, Rebecca LS, et al. Global cancer statistics 2020: GLOBOCAN estimates of incidence and mortality worldwide for 36 cancers in 185 countries. CA: A Cancer Journal for Clinicians. 2021;**71**(3):209-249. DOI: 10.3322/caac.21660

[3] Siegel RL, Giaquinto AN, Jemal A. Cancer statistics, 2024. CA: A Cancer Journal for Clinicians. 2024;**74**(1):12-49. DOI: 10.3322/caac.21820

[4] Zhang ZH, Wang Y, Zhang Y, et al. The function and mechanisms of action of circular RNAs in urologic cancer. Molecular Cancer. 2023;**22**(1):61. DOI: 10.1186/s12943-023-01766-2

[5] Shahneen S, Caroline MM, Edmund C, Himisha B, Robert GB, Scott W. Prostate cancer. The Lancet. 2021;**398**(10305):1075-1090. DOI: 10.1016/s0140-6736(21)00950-8

[6] Regan MM, Mantia CM, Werner L, et al. Treatment-free survival over extended follow-up of patients with advanced melanoma treated with immune checkpoint inhibitors in CheckMate 067. Journal for Immunotherapy of Cancer. 2021;**9**(11):e003743-e003743. DOI: 10.1136/jitc-2021-003743

[7] Wei G, Sun H, Dong K, et al. The thermogenic activity of adjacent adipocytes fuels the progression of ccRCC and compromises anti-tumor therapeutic efficacy. Cell Metabolism. 2021;**33**(10):2021-2039.e8. DOI: 10.1016/j.cmet.2021.08.012

[8] Nadal R, Bellmunt J. Management of metastatic bladder cancer. Cancer Treatment Reviews. 2019;**76**:10-21. DOI: 10.1016/j.ctrv.2019.04.002

[9] Nair SS, Chakravarty D, Patel V, Bhardwaj N, Tewari AK. Genitourinary cancer neoadjuvant therapies: Current and future approaches. Trends Cancer. 2023;**9**(12):1041-1057. DOI: 10.1016/j.trecan.2023.07.011

[10] Mauricio R-M, Gabriel RC, Gabriel M-R, Rodrigo APM. From gene targeting to genome editing: Transgenic animals applications and beyond. Anais Da Academia Brasileira De Ciencias. 2015;**87**(2 suppl):1323-1348. DOI: 10.1590/0001-3765201520140710

[11] Hao Y, Kevin KJ, Anderson DG. Delivery technologies for genome editing. Nature Reviews. Drug Discovery. 2017;**16**(6):387-399. DOI: 10.1038/nrd.2016.280

[12] Jun-Gyu P, Paula AP, Anwari A, Xavier A, Jordi BT, Luis M-S. Animal models of COVID-19: Transgenic mouse model. Methods in Molecular Biology. 2022:259-289. DOI: 10.1007/978-1-0716-2111-0_16

[13] Laszlo D, Katalin Ö, Orsolya H, et al. Characterization of cardiac effects of OPA1 protein promotion in transgenic animal model. European Heart Journal. 2022;**43**(Supplement_2). DOI: 10.1093/eurheartj/ehac544.2892

[14] Amélie C, Charbel C, Silvia S, et al. Tu1881 HLA B27 transgenic rat: A new animal model of postsurgical ileitis in inflammatory bowel disease. Gastroenterology. 2016;**150**(4):S967-S967. DOI: 10.1016/s0016-5085(16)33275-9

[15] Sambhavi B, Amulya V, Anandan B. Genetically engineered and spontaneous animal model: Utilization in preclinical cancer therapy development. In: Handbook of Animal Models and Its Uses in Cancer Research. 2023. pp. 421-431. DOI: 10.1007/978-981-19-1282-5_22-1

[16] Sousa ER, Zoni E, Scarpa M, et al. Cripto as a key factor in aggressive lethal prostate cancer: Generation of a new transgenic mouse model. European Urology. 2023;**83**:S1667-S1667. DOI: 10.1016/s0302-2838(23)01192-2

[17] Aylin A-D, James LM, Erin M, Justyna G, Han WT, John AC. Abstract 3041: ASPSCR1-TFE3 gene fusion in transgenic mouse as a new pediatric kidney cancer model. Tumor Biology. 2021;**81**(13_Supplement):3041-3041. DOI: 10.1158/1538-7445.am2021-3041

[18] Aysegul B, Shinjini G, Petros G, et al. A novel orthotopic murine model of neuroendocrine bladder cancer: Insights into the phenotypic plasticity of small cell bladder cancer (SCBC). Journal of Clinical Oncology. 2020;**38**(6_suppl):571-571. DOI: 10.1200/jco.2020.38.6_suppl.571

[19] Norman MG, Francesco JD, Milton JF, et al. Prostate cancer in a transgenic mouse. Proceedings of the National Academy of Sciences. 1995;**92**(8):3439-3443. DOI: 10.1073/pnas.92.8.3439

[20] Larissa Akemi K, de Almeida Lamas C, Marostica MR, Cagnon VHA. Transgenic adenocarcinoma of the mouse prostate (TRAMP) model: A good alternative to study PCa progression and chemoprevention approaches. Life Sciences. 2019;**217**:141-147. DOI: 10.1016/j.lfs.2018.12.002

[21] Irwin HG. How the TRAMP model revolutionized the study of prostate cancer progression. Cancer Research. 2016;**76**(21):6137-6139. DOI: 10.1158/0008-5472.Can-16-2636

[22] Arthur AH, Foster BA, Allison JP, Greenberg NM, Kwon ED. The TRAMP mouse as a model for prostate cancer. Current Protocols in Immunology. 2001;**Chapter 20**:20.5.1-20.5.23. DOI: 10.1002/0471142735.im2005s45

[23] Lisa DB-B, Karen EK. Models of neuroendocrine prostate cancer. Endocrine-Related Cancer. 2014;**22**(1):R33-R49. DOI: 10.1530/erc-14-0393

[24] Chen R, Liang X, Murray MM, et al. A simple quantitative PCR assay to determine TRAMP transgene zygosity. Prostate Cancer and Prostatic Diseases. 2021;**24**(2):358-361. DOI: 10.1038/s41391-020-00282-4

[25] Jeffrey RG, Roberto B, Ronald AM, et al. Metastatic prostate cancer in a transgenic mouse. Cancer Research. 1996;**56**(18):4096-4102

[26] Teresa TL, Jonathan AE, Emily AR, Robert HB, Colin CC, William AR. Modeling human prostate cancer progression in vitro. Carcinogenesis. 2018;**40**(7):893-902. DOI: 10.1093/carcin/bgy185

[27] Jinny S, Robert B, Justin Delos S, et al. Resistance to androgen deprivation leads to altered metabolism in human and murine prostate cancer cell and tumor models. Metabolites. 2021;**11**(3):139-139. DOI: 10.3390/metabo11030139

[28] Susan K, Patricia S, Yuhao Y, et al. Development, progression, and androgen-dependence of prostate tumors in probasin-large T antigen transgenic mice: A model for prostate cancer. Laboratory Investigation. A Journal

of Technical Methods and Pathology. 1998;**78**(6):i-xv

[29] Olga K, John RC, Janni M, Richard LR, Robert JM, Valeri V. Hepsin promotes prostate cancer progression and metastasis. Cancer Cell. 2004;**6**(2):185-195. DOI: 10.1016/j.ccr.2004.07.008

[30] Sirin S, Sachin Kumar D, Santanu D, Sachin P, Seema S, Ajay PS. Cellular and molecular progression of prostate cancer: Models for basic and preclinical research. Cancers. 2020;**12**(9):2651-2651. DOI: 10.3390/cancers12092651

[31] Arriaga JM, Abate-Shen C. Genetically engineered mouse models of prostate cancer in the postgenomic era. Cold Spring Harbor Perspectives in Medicine. 2018;**9**(2):a030528-a030528. DOI: 10.1101/cshperspect.a030528

[32] Susan K. Survey of genetically engineered mouse models for prostate cancer: Analyzing the molecular basis of prostate cancer development, progression, and metastasis. Journal of Cellular Biochemistry. 2005;**94**(2):279-297. DOI: 10.1002/jcb.20339

[33] Maroulakou IG, Anver M, Garrett L, Green JE. Prostate and mammary adenocarcinoma in transgenic mice carrying a rat C3(1) simian virus 40 large tumor antigen fusion gene. Proceedings of the National Academy of Sciences of the United States of America. 1994;**91**(23):11236-11240. DOI: 10.1073/pnas.91.23.11236

[34] Ingrid E, Michelle Y, Jennifer W, et al. A low-carbohydrate diet containing soy protein and fish oil reduces breast but not prostate cancer in C3(1)/tag mice. Carcinogenesis. 2021;**43**(2):115-125. DOI: 10.1093/carcin/bgab106

[35] Nancy EM, Jennifer MT-A, Lei W, Krystle EZ, John WE, Steven KC. Tomatoes, lycopene, and prostate cancer: What have we learned from experimental models? Journal of Nutrition. 2022;**152**(6):1381-1403. DOI: 10.1093/jn/nxac066

[36] Ioanna GM, Miriam RA, Lisa G, Jeffrey EG. Prostate and mammary adenocarcinoma in transgenic mice carrying a rat C3(1) simian virus 40 large tumor antigen fusion gene. Proceedings of the National Academy of Sciences of the United States of America. 1994;**91**(23):11236-11240. DOI: 10.1073/pnas.91.23.11236

[37] Wenhao Z, Wytske M v W, Corrina MAdR, et al. Ex vivo treatment of prostate tumor tissue recapitulates in vivo therapy response. The Prostate. 2018;**79**(4):390-402. DOI: 10.1002/pros.23745

[38] Emily G, Peter AH, Jeffrey IG. A transgenic mouse model of metastatic prostate cancer originating from neuroendocrine cells. Proceedings of the National Academy of Sciences of the United States of America. 1998;**95**(26):15382-15387. DOI: 10.1073/pnas.95.26.15382

[39] Kosuke O, Kei M, Tomohiro F, et al. Establishment and characterization of a novel treatment-related neuroendocrine prostate cancer cell line KUCaP13. Cancer Science. 2021;**112**(7):2781-2791. DOI: 10.1111/cas.14935

[40] Sedlack AJH, Saleh-Anaraki K, Kumar S, et al. Preclinical models of neuroendocrine neoplasia. Cancers (Basel). 2022;**14**:22. DOI: 10.3390/cancers14225646

[41] Tarana A, Peter SN. Endocrine and paracrine characteristics of neuroendocrine prostate cancer.

Frontiers in Endocrinology. 2022;**13**. DOI: 10.3389/fendo.2022.1012005

[42] Wenming D, Manal G, Madeleine M, et al. Knockin of SV40 tag oncogene in a mouse adenocarcinoma of the prostate model demonstrates advantageous features over the transgenic model. Oncogene. 2005;**24**(9):1510-1524. DOI: 10.1038/sj.onc.1208229

[43] Manal G, Wenming D, Guojun W, et al. A novel knock-in prostate cancer model demonstrates biology similar to that of human prostate cancer and suitable for preclinical studies. Molecular Therapy. 2005;**11**(3):348-362. DOI: 10.1016/j.ymthe.2004.12.005

[44] Wu G, Yu L-C, Wang L, et al. Application of Gleason analogous grading system and flow cytometry DNA analysis in a novel knock-in mouse prostate cancer model. Postgraduate Medical Journal. 2006;**82**(963):40-45. DOI: 10.1136/pgmj.2005.038042

[45] Lee YR, Chen M, Pandolfi PP. The functions and regulation of the PTEN tumour suppressor: New modes and prospects. Nature Reviews. Molecular Cell Biology. 2018;**19**(9):547-562. DOI: 10.1038/s41580-018-0015-0

[46] Maisa Y, Cutz JC, Nuin PA, et al. Interphase FISH analysis of PTEN in histologic sections shows genomic deletions in 68% of primary prostate cancer and 23% of high-grade prostatic intra-epithelial neoplasias. Cancer Genetics and Cytogenetics. 2006;**169**(2):128-137. DOI: 10.1016/j.cancergencyto.2006.04.003

[47] Holcomb IN, Young JM, Coleman IM, et al. Comparative analyses of chromosome alterations in soft-tissue metastases within and across patients with castration-resistant prostate cancer. Cancer Research. 2009;**69**(19):7793-7802. DOI: 10.1158/0008-5472.Can-08-3810

[48] Shunyou W, Gao J, Lei Q, et al. Prostate-specific deletion of the murine Pten tumor suppressor gene leads to metastatic prostate cancer. Cancer Cell. 2003;**4**(3):209-221. DOI: 10.1016/s1535-6108(03)00215-0

[49] Di Cristofano A, Pesce B, Cordon-Cardo C, Pandolfi PP. Pten is essential for embryonic development and tumour suppression. Nature Genetics. 1998;**19**(4):348-355. DOI: 10.1038/1235

[50] Podsypanina K, Ellenson LH, Nemes A, et al. Mutation of Pten/Mmac1 in mice causes neoplasia in multiple organ systems. Proceedings of the National Academy of Sciences of the United States of America. 1999;**96**(4):1563-1568. DOI: 10.1073/pnas.96.4.1563

[51] Stambolic V, Tsao MS, Macpherson D, Suzuki A, Chapman WB, Mak TW. High incidence of breast and endometrial neoplasia resembling human Cowden syndrome in pten+/− mice. Cancer Research. 2000;**60**(13):3605-3611

[52] Qingli L, Lunxu L, Zhang Q, Len S, Ge D, You Z. Interleukin-17 indirectly promotes M2 macrophage differentiation through stimulation of COX-2/PGE2 pathway in the cancer cells. Cancer Research and Treatment. 2014;**46**(3):297-306. DOI: 10.4143/crt.2014.46.3.297

[53] Qiuyang Z, Liu S, Zhang Q, et al. Interleukin-17 promotes development of castration-resistant prostate cancer potentially through creating an immunotolerant and pro-angiogenic tumor microenvironment. The Prostate. 2014;**74**(8):869-879. DOI: 10.1002/pros.22805

[54] Feng B, Pei XH, Pandolfi PP, Xiong Y. p18 Ink4c and Pten constrain

a positive regulatory loop between cell growth and cell cycle control. Molecular and Cellular Biology. 2006;**26**(12):4564-4576. DOI: 10.1128/mcb.00266-06

[55] Kim MJ, Cardiff RD, Desai N, et al. Cooperativity of Nkx3.1 and Pten loss of function in a mouse model of prostate carcinogenesis. Proceedings of the National Academy of Sciences of the United States of America. 2002;**99**(5):2884-2889. DOI: 10.1073/pnas.042688999

[56] Alvaro A, Antonina M, Carolyn Waugh K, et al. ETV4 promotes metastasis in response to activation of PI3-kinase and Ras signaling in a mouse model of advanced prostate cancer. Proceedings of the National Academy of Sciences. 2013;**110**(37). DOI: 10.1073/pnas.1303558110

[57] Zhihu D, Chang Jiun W, Gerald CC, et al. SMAD4-dependent barrier constrains prostate cancer growth and metastatic progression. Nature. 2011;**470**(7333):269-273. DOI: 10.1038/nature09677

[58] Jayaprakash P, Ai M, Liu A, et al. Targeted hypoxia reduction restores T cell infiltration and sensitizes prostate cancer to immunotherapy. The Journal of Clinical Investigation. 2018;**128**(11):5137-5149. DOI: 10.1172/jci96268

[59] Yini Z, Zhao Y, Wen J, et al. Targeting the chromatin effector Pygo2 promotes cytotoxic T cell responses and overcomes immunotherapy resistance in prostate cancer. Science Immunology. 2023;**8**(81):eade4656. DOI: 10.1126/sciimmunol.ade4656

[60] Mirjam B, Liu D, Brian DR, et al. SPOP mutation drives prostate tumorigenesis in vivo through coordinate regulation of PI3K/mTOR and AR signaling. Cancer Cell. 2017;**31**(3):436-451. DOI: 10.1016/j.ccell.2017.02.004

[61] David JM, Kobayashi N, Marcus R, et al. Pten loss and RAS/MAPK activation cooperate to promote EMT and metastasis initiated from prostate cancer stem/progenitor cells. Cancer Research. 2012;**72**(7):1878-1889. DOI: 10.1158/0008-5472.can-11-3132

[62] Bora G, Takashi I, Cheryl MK, et al. Nuclear MYC protein overexpression is an early alteration in human prostate carcinogenesis. Modern Pathology. 2008;**21**(9):1156-1167. DOI: 10.1038/modpathol.2008.111

[63] Cheryl MK, Charles JB, Chi VD, William GN, Srinivasan Y, Angelo MDM. MYC and prostate cancer. Genes & Cancer. 2010;**1**(6):617-628. DOI: 10.1177/1947601910379132

[64] Katharine E-Y, Thomas GG, John W, et al. Myc-driven murine prostate cancer shares molecular features with human prostate tumors. Cancer Cell. 2005;**8**(6):485-485. DOI: 10.1016/s1535-6108(03)00197-1

[65] Gretchen KH, Laura NM, May K, et al. Combined MYC activation and Pten loss are sufficient to create genomic instability and lethal metastatic prostate cancer. Cancer Research. 2016;**76**(2):283-292. DOI: 10.1158/0008-5472.can-14-3280

[66] McMullin RP, Mutton LN, Bieberich CJ. Hoxb13 regulatory elements mediate transgene expression during prostate organogenesis and carcinogenesis. Developmental Dynamics. 2009;**238**(3):664-672. DOI: 10.1002/dvdy.21870

[67] Srinivas N, Katharine E-Y, Charles LS, et al. Hepsin cooperates with MYC in the progression of

adenocarcinoma in a prostate cancer mouse model. The Prostate. 2009;**70**(6):591-600. DOI: 10.1002/pros.21093

[68] Clegg NJ, Couto SS, Wongvipat J, et al. MYC cooperates with AKT in prostate tumorigenesis and alters sensitivity to mTOR inhibitors. PLoS One. 2011;**6**(3):e17449. DOI: 10.1371/journal.pone.0017449

[69] William HF, Andre H, Robert DM, et al. Expression of the c-myc protooncogene in human prostatic carcinoma and benign prostatic hyperplasia. Cancer Research. 1986;**46**(3):1535-1538

[70] Paul CB, Michael F, Nicholas JH, et al. Spatial genomic heterogeneity within localized, multifocal prostate cancer. Nature Genetics. 2015;**47**(7):736-745. DOI: 10.1038/ng.3315

[71] Himisha B, Davide P, Juan Miguel M, et al. Divergent clonal evolution of castration-resistant neuroendocrine prostate cancer. Nature Medicine. 2016;**22**(3):298-305. DOI: 10.1038/nm.4045

[72] Etienne D, Himisha B, Matteo B, et al. N-Myc induces an EZH2-mediated transcriptional program driving neuroendocrine prostate cancer. Cancer Cell. 2016;**30**(4):563-577. DOI: 10.1016/j.ccell.2016.09.005

[73] Abdullah A, Mariantonietta DA, Martina T, et al. CDCP1 overexpression drives prostate cancer progression and can be targeted in vivo. Journal of Clinical Investigation. 2020;**130**(5):2435-2450. DOI: 10.1172/jci131133

[74] Kuo CY, Hsieh PC, Chiu V, Lan CC, Lu KC. The von Hippel-Lindau tumor suppressor gene mutations modulate Lipocalin-2 expression in Ferroptotic-inflammatory pathways. Oxidative Medicine and Cellular Longevity. 2023;**2023**:7736638. DOI: 10.1155/2023/7736638

[75] Marco G, Stuart H, James L, et al. Genomic architecture and evolution of clear cell renal cell carcinomas defined by multiregion sequencing. Nature Genetics. 2014;**46**(3):225-233. DOI: 10.1038/ng.2891

[76] Yusuke S, Tetsuichi Y, Yuichi S, et al. Integrated molecular analysis of clear-cell renal cell carcinoma. Nature Genetics. 2013;**45**(8):860-867. DOI: 10.1038/ng.2699

[77] Erinn BR, John ET, Volker HH. Renal cyst development in mice with conditional inactivation of the von Hippel-Lindau tumor suppressor. Cancer Research. 2006;**66**(5):2576-2583. DOI: 10.1158/0008-5472.can-05-3241

[78] Ian JF, Claudio RT, Strahil G, et al. pVHL and PTEN tumour suppressor proteins cooperatively suppress kidney cyst formation. The EMBO Journal. 2008;**27**(12):1747-1757. DOI: 10.1038/emboj.2008.96

[79] Gunnar S, Bernd K, Johannes S, et al. Hypoxia-inducible transcription factors stabilization in the thick ascending limb protects against ischemic acute kidney injury. Journal of the American Society of Nephrology. 2011;**22**(11):2004-2015. DOI: 10.1681/asn.2010121249

[80] Ruth S, Thomas H, Maxine T, et al. Renal tubular HIF-2α expression requires VHL inactivation and causes fibrosis and cysts. PLoS One. 2012;**7**(1):e31034-e31034. DOI: 10.1371/journal.pone.0031034

[81] Tracy LP, Hannah LB, Joel MH, Tien H. Conditional inactivation of the mouse von Hippel–Lindau tumor

suppressor gene results in wide-spread hyperplastic, inflammatory and fibrotic lesions in the kidney. Oncogene. 2014;**34**(20):2631-2639. DOI: 10.1038/onc.2014.197

[82] Rafał P, Sarah MM, Tullio S, Wilhelm K, Holger M, Peter S. Loss of PBRM1 expression is associated with renal cell carcinoma progression. International Journal of Cancer. 2012;**132**(2). DOI: 10.1002/ijc.27822

[83] Ignacio V, Patrick T, Keiran R, et al. Exome sequencing identifies frequent mutation of the SWI/SNF complex gene PBRM1 in renal carcinoma. Nature. 2011;**469**(7331):539-542. DOI: 10.1038/nature09639

[84] Xun X, Yong H, Xue Y, et al. Single-cell exome sequencing reveals single-nucleotide mutation characteristics of a kidney tumor. Cell. 2012;**148**(5):886-895. DOI: 10.1016/j.cell.2012.02.025

[85] Lucy G, Muhammed M, Andrew FS, et al. Clinical and pathological impact of VHL, PBRM1, BAP1, SETD2, KDM6A, and JARID1cin clear cell renal cell carcinoma. Genes, Chromosomes and Cancer. 2013;**53**(1):38-51. DOI: 10.1002/gcc.22116

[86] Jeanette EE-P, Daniel JS, John CC, et al. BAP1 and PBRM1 in metastatic clear cell renal cell carcinoma: Tumor heterogeneity and concordance with paired primary tumor. BMC Urology. 2017;**17**(1). DOI: 10.1186/s12894-017-0209-3

[87] Amrita MN, Can GP, Yiyu D, et al. The SWI/SNF protein PBRM1 restrains VHL-loss-driven clear cell renal cell carcinoma. Cell Reports. 2017;**18**(12):2893-2906. DOI: 10.1016/j.celrep.2017.02.074

[88] Yi G, Shannon MC, Alana C, et al. Modeling renal cell carcinoma in mice: Bap1 and Pbrm1 inactivation drive tumor grade. Cancer Discovery. 2017;**7**(8):900-917. DOI: 10.1158/2159-8290.cd-17-0292

[89] Shan, Shan W, Yi G, Nicholas CW, et al. Bap1 is essential for kidney function and cooperates with Vhl in renal tumorigenesis. Proceedings of the National Academy of Sciences. 2014;**111**(46):16538-16543. DOI: 10.1073/pnas.1414789111

[90] James B. Molecular genetics of clear-cell renal cell carcinoma. Journal of Clinical Oncology. 2014;**32**(18):1968-1976. DOI: 10.1200/jco.2012.45.2003

[91] Sabine H, Désirée S, Nora CT, et al. Combined mutation in Vhl, Trp53 and Rb1 causes clear cell renal cell carcinoma in mice. Nature Medicine. 2017;**23**(7):869-877. DOI: 10.1038/nm.4343

[92] Sean TB, Aleisha MS, Jordan K, et al. MYC activation cooperates with Vhl and Ink4a/Arf loss to induce clear cell renal cell carcinoma. Nature Communications. 2017;**8**(1). DOI: 10.1038/ncomms15770

[93] Michael LN, Michelle BW, Jorge RT, et al. Mutations in a novel gene lead to kidney tumors, lung wall defects, and benign tumors of the hair follicle in patients with the Birt-Hogg-Dubé syndrome. Cancer Cell. 2002;**2**(2):157-164. DOI: 10.1016/s1535-6108(02)00104-6

[94] Birt AR. Hereditary multiple fibrofolliculomas with trichodiscomas and acrochordons. Archives of Dermatology. 1977;**113**(12):1674-1677. DOI: 10.1001/archderm.113.12.1674

[95] Tiffiney RH, Emmanuelle N, Andres JK-S, et al. The role of the Birt–Hogg–Dubé protein in mTOR activation and renal tumorigenesis. Oncogene. 2009;**28**(13):1594-1604. DOI: 10.1038/onc.2009.14

[96] Yukiko H, Baba M, Rieko A, et al. Homozygous loss of BHD causes early embryonic lethality and kidney tumor development with activation of mTORC1 and mTORC2. Proceedings of the National Academy of Sciences of the United States of America. 2009;**106**(44):18722-18727. DOI: 10.1073/pnas.0908853106

[97] Valérie H, Sylvie S, Anders Bondo D, et al. Renal tumour suppressor function of the Birt-Hogg-Dube syndrome gene product folliculin. Journal of Medical Genetics. 2009;**47**(3):182-189. DOI: 10.1136/jmg.2009.072009

[98] Jindong C, Kunihiko F, David P, et al. Deficiency of FLCN in mouse kidney led to development of polycystic kidneys and renal neoplasia. PLoS One. 2008;**3**(10):e3581-e3581. DOI: 10.1371/journal.pone.0003581

[99] Mingsong W, Shuhui S, Yan L, et al. Flcn-deficient renal cells are tumorigenic and sensitive to mTOR suppression. Oncotarget. 2015;**6**(32):32761-32773. DOI: 10.18632/oncotarget.5018

[100] Jindong C, Dachuan H, Isabelle R, et al. Disruption of tubular Flcn expression as a mouse model for renal tumor induction. Kidney International. 2015;**88**(5):1057-1069. DOI: 10.1038/ki.2015.177

[101] Zachary SM, Andrea IM. Aberrant epithelial morphology and persistent epidermal growth factor receptor signaling in a mouse model of renal carcinoma. Proceedings of the National Academy of Sciences. 2009;**106**(24):9767-9772. DOI: 10.1073/pnas.0902031106

[102] Leiping F, Gang W, Maria MS, David MN, Lorraine JG. Generation of a mouse model of Von Hippel–Lindau kidney disease leading to renal cancers by expression of a constitutively active mutant of HIF1α. Cancer Research. 2011;**71**(21):6848-6856. DOI: 10.1158/0008-5472.can-11-1745

[103] Alessia C, Lotte K, Eric HJV, et al. Modelling TFE renal cell carcinoma in mice reveals a critical role of WNT signaling. eLife. 2016;**5**. DOI: 10.7554/elife.17047

[104] Qianghua H, Fei G, Weihua T, et al. Wt1 ablation and Igf2 upregulation in mice result in Wilms tumors with elevated ERK1/2 phosphorylation. Journal of Clinical Investigation. 2011;**121**(1):174-183. DOI: 10.1172/jci43772

[105] Leo GF, Hung-Chieh Y, Suren S, et al. Monitoring therapy with MEK inhibitor U0126 in a novel Wilms tumor model in Wt1 knockout Igf2 transgenic mice using 18F-FDG PET with dual-contrast enhanced CT and MRI: Early metabolic response without inhibition of tumor growth. Molecular Imaging and Biology. 2012;**15**(2):175-185. DOI: 10.1007/s11307-012-0588-5

[106] Wu XR. Biology of urothelial tumorigenesis: Insights from genetically engineered mice. Cancer Metastasis Reviews. 2009;**28**(3-4):281-290. DOI: 10.1007/s10555-009-9189-4

[107] Regina A-R, Aura C, Rosário P-L, Paula AO. In vitro and in vivo experimental models as tools to investigate the efficacy of antineoplastic drugs on urinary bladder cancer. Anticancer Research. 2013;**33**(4):1273-1296

[108] Zhong-Ting Z, Joanne P, Ellen S, Tung-Tien S, Xue-Ru W. Urothelium-specific expression of an oncogene in transgenic mice induced the formation of carcinoma in situ and invasive

transitional cell carcinoma. Cancer Research. 1999;**59**(14):3512-3517

[109] Ayala de la Peña F, Kanasaki K, Kanasaki M, Tangirala N, Maeda G, Kalluri R. Loss of p53 and acquisition of angiogenic microRNA profile are insufficient to facilitate progression of bladder urothelial carcinoma in situ to invasive carcinoma. The Journal of Biological Chemistry. 2011;**286**(23):20778-20787. DOI: 10.1074/jbc.M110.198069

[110] Reddy EP, Roberta KR, Eugenio S, Mariano B. A point mutation is responsible for the acquisition of transforming properties by the T24 human bladder carcinoma oncogene. Nature. 1982;**300**(5888):149-152. DOI: 10.1038/300149a0

[111] Zhong, Ting Z, Joanne P, Hong H, et al. Role of Ha-ras activation in superficial papillary pathway of urothelial tumor formation. Oncogene. 2001;**20**(16):1973-1980. DOI: 10.1038/sj.onc.1204315

[112] Arnold JL. p53, the cellular gatekeeper for growth and division. Cell. 1997;**88**(3):323-331. DOI: 10.1016/s0092-8674(00)81871-1

[113] Xue-Ru W. Urothelial tumorigenesis: A tale of divergent pathways. Nature Reviews Cancer. 2005;**5**(9):713-725. DOI: 10.1038/nrc1697

[114] Jing G, Hong H, Joanne P, et al. p53 deficiency provokes urothelial proliferation and synergizes with activated Ha-ras in promoting urothelial tumorigenesis. Oncogene. 2004;**23**(3):687-696. DOI: 10.1038/sj.onc.1207169

[115] Anna MP-K, Mireia C-M, Carolyn Waugh K, et al. Inactivation of p53 and Pten promotes invasive bladder cancer. Genes & Development. 2009;**23**(6):675-680. DOI: 10.1101/gad.1772909

[116] Chao-Nan Q, Kyle AF, Jared K, et al. Activation of the PI3K/AKT pathway induces urothelial carcinoma of the renal pelvis: Identification in human tumors and confirmation in animal models. Cancer Research. 2009;**69**(21):8256-8264. DOI: 10.1158/0008-5472.can-09-1689

[117] Kenneth AI. Germ cell neoplasms of the testis: Update for 2022. Seminars in Diagnostic Pathology. 2023;**40**(1):2-21. DOI: 10.1053/j.semdp.2022.07.001

[118] Gill JA, Lowe L, Nguyen J, et al. Enforced expression of simian virus 40 large T-antigen leads to testicular germ cell tumors in zebrafish. Zebrafish. 2010;**7**(4):333-341. DOI: 10.1089/zeb.2010.0663

[119] Leroy CS, Little CC. Spontaneous testicular teratomas in an inbred strain of mice. Proceedings of the National Academy of Sciences of the United States of America. 1954;**40**(11):1080-1087. DOI: 10.1073/pnas.40.11.1080

[120] Leroy CS. Studies on transplantable testicular teratomas of strain 129 Mice2. Journal of the National Cancer Institute. 1958;**20**(6):1257-1275. DOI: 10.1093/jnci/20.6.1257

[121] Leroy CS. A new inbred subline of mice (129/terSv) with a high incidence of spontaneous congenital testicular Teratomas2. Journal of the National Cancer Institute. 1973;**50**(1):235-242. DOI: 10.1093/jnci/50.1.235

[122] Timothy MP, Amy ML, Claire A, et al. Chemotherapy-induced depletion of OCT4-positive cancer stem cells in a mouse model of malignant testicular cancer. Cell Reports.

2017;**21**(7):1896-1909. DOI: 10.1016/j.celrep.2017.10.078

[123] Lyndaker AM, Pierpont TM, Loehr AR, Weiss RS. A genetically engineered mouse model of malignant testicular germ cell tumors. Methods in Molecular Biology. 2021;**2195**:147-165. DOI: 10.1007/978-1-0716-0860-9_11

[124] Antoin D, Timothy AM. Updates on the epidemiology and risk factors for penile cancer. Translational Andrology and Urology. 2017;**6**(5):785-790. DOI: 10.21037/tau.2017.05.19

[125] Catherine, de M, Martyn P, Jerome V, Silvia F. Worldwide burden of cancer attributable to HPV by site, country and HPV type. International Journal of Cancer. 2017;**141**(4):664-670. DOI: 10.1002/ijc.30716

[126] Beatriz M-F, Verónica FM, Diogo E, et al. HPV16induces penile intraepithelial neoplasia and squamous cell carcinoma in transgenic mice: First mouse model for HPV-related penile cancer. The Journal of Pathology. 2020;**251**(4):411-419. DOI: 10.1002/path.5475

[127] Martin AN, Frederick MA. Introduction to gene editing and manipulation using CRISPR/Cas9 technology. Current Protocols in Molecular Biology. 2016;**115**(1). DOI: 10.1002/cpmb.14

[128] Hino T, Omura SN, Nakagawa R, et al. An AsCas12f-based compact genome-editing tool derived by deep mutational scanning and structural analysis. Cell. 2023;**186**(22):4920-4935 e23. DOI: 10.1016/j.cell.2023.08.031

[129] Akiko T, Hiroyuki S, Tomoo O, et al. Inducing multiple nicks promotes interhomolog homologous recombination to correct heterozygous mutations in somatic cells. Nature Communications. 2023;**14**(1). DOI: 10.1038/s41467-023-41048-5

[130] Minjung K, Rajula B-G, Whitney B-P, et al. Nkx3.1 mutant mice recapitulate early stages of prostate carcinogenesis. Cancer Research. 2002;**62**(11):2999-3004

Chapter 7

Gene Therapy for Hypophosphatasia: Current Management and Future

Tae Matsumoto, Sonoko Narisawa, José Luis Millán and Koichi Miyake

Abstract

This review provides a comprehensive overview of hypophosphatasia (HPP), a rare genetic disorder marked by defective bone and teeth mineralization due to mutations in the *ALPL* gene. It reviews the evolution of HPP treatment, from early symptomatic management methods to the latest therapeutic approaches, emphasizing significant milestones achieved over time. In addition, this review delves into gene therapy's historical development, outlining its successes and challenges. Furthermore, it presents a detailed analysis of why this innovative therapy holds promise for HPP, focusing on its efficacy in correcting the underlying biochemical defects and significantly improving patient outcomes. Moreover, the review discusses future research directions, highlighting the critical need for ongoing innovation and rigorous clinical trials to further enhance the efficacy and safety of gene therapy for HPP. Continuous research is essential to developing more effective treatments and ultimately improving the quality of life for patients affected by this debilitating condition.

Keywords: Hypophosphatasia, alkaline phosphatase, therapeutics, progress, genetic therapies, adeno-associated virus vector

1. Introduction

Hypophosphatasia (HPP) is a rare genetic disorder characterized by defective mineralization of bones and teeth due to low activity of the tissue-nonspecific alkaline phosphatase (TNAP) isozyme, encoded by the *ALPL* gene. The severe forms of HPP primarily exhibit autosomal recessive inheritance, whereas the milder forms predominantly exhibit autosomal dominant inheritance. Symptoms vary widely from embryonic to adult onset, depending on the levels of residual alkaline phosphatase (ALP) activity. The disorder is classified into six types: perinatal, benign prenatal, infantile, childhood, adult, and odontohypophosphatasia [1, 2].

Inorganic pyrophosphate (PPi), a calcification inhibitor, is typically hydrolyzed by TNAP. Elevated PPi levels in the plasma of patients with HPP prevent hydroxyapatite mineral precipitation and growth. Another substrate of TNAP is pyridoxal-5'-phosphate (PLP), which is essential for a wide range of biochemical reactions,

 IntechOpen

including neurotransmitter synthesis [3]; its deficiency in the brain can cause vitamin B6-dependent convulsions [2]. In addition, urinary phosphoethanolamine (PEA) levels are elevated in patients with HPP, aiding in the diagnosis of HPP. However, the clinical significance of PEA levels and the affected metabolic pathways remain unclear.

The incidence of severe HPP is reported to be 1/100,000 people in North America [4], 1/300,000 people in Europe [5], and 2–3/1,000,000 people in Japan [6, 7], whereas the nonlethal forms are more prevalent. Actual prevalence of HPP is possibly higher than previously reported, as clinical symptoms are less prominent in certain individuals, while the same mutations cause severe phenotype in other patients. Additionally, many patients with minor illnesses may remain undiagnosed [8, 9]. For example, the del1559 mutation in *ALPL* is frequently observed in Japanese population, and its carrier incidence is approximately 1 in 480 individuals [8]. Adult-onset HPP often presents with nonspecific symptoms, such as chronic pain and osteoporosis-like radiographic findings, leading to frequent misdiagnosis and may go undetected because of inaccurate diagnosis [10–12]. The number of patients with HPP may be higher than previously reported, as some cases are asymptomatic but have genetic mutations in TNAP or heterozygous genetic mutations [8, 13]. Furthermore, over 400 mutations of the *ALPL* gene have been reported [14, 15]; these mutations result in a spectrum of clinical manifestations, ranging from mild to severe skeletal abnormalities.

As a single-enzyme-deficient disease, HPP is suitable for therapeutic interventions. Asfotase alfa, a recombinant enzyme replacement therapy (ERT), became available in 2015 [16] and revolutionized HPP treatment. However, ERT requires frequent injections throughout a patient's life, posing a burden, particularly in pediatric cases with early onset of the disease [17]. The success of ERT has propelled gene therapy to become the next frontier in HPP treatment. In this chapter, we review the history of HPP treatment, evolution of gene therapy, ongoing gene therapy research on HPP, and prospects for future treatments.

2. Pathophysiology

In children with HPP, serum ALP levels are significantly lower than age-matched normal levels. The reference range for serum ALP varies with age and is typically higher in children due to active bone metabolism during growth [18]. Thus, ALP values that are usually normal in adults may fall below the normal range in children, often leading to misinterpretation and delays in diagnosis and treatment. Therefore, age and sex must be considered when assessing ALP levels to avoid overlooking HPP cases [19].

Furthermore, HPP in pregnancy requires caution. Owing to the presence of placenta-derived ALP, serum ALP activity is generally higher in pregnant than in nonpregnant women. Thus, low plasma ALP activity could be normal or high during pregnancy, masking a physiologically low bone-derived TNAP [20, 21].

The *ALPL* gene, recently renamed "alkaline phosphatase, biomineralization associated" [22], encodes TNAP and is located at 1p36.12 on the short arm of chromosome 1. More than 400 mutations have been reported associated with HPP [14, 15], and the genotype and phenotype do not always correlate. Patients with identical genotypes may exhibit variations in the timing of disease onset and symptom presentation. Predicting symptom onset and presentation in later-born children, even if they share the same genetic mutation as earlier-born children in

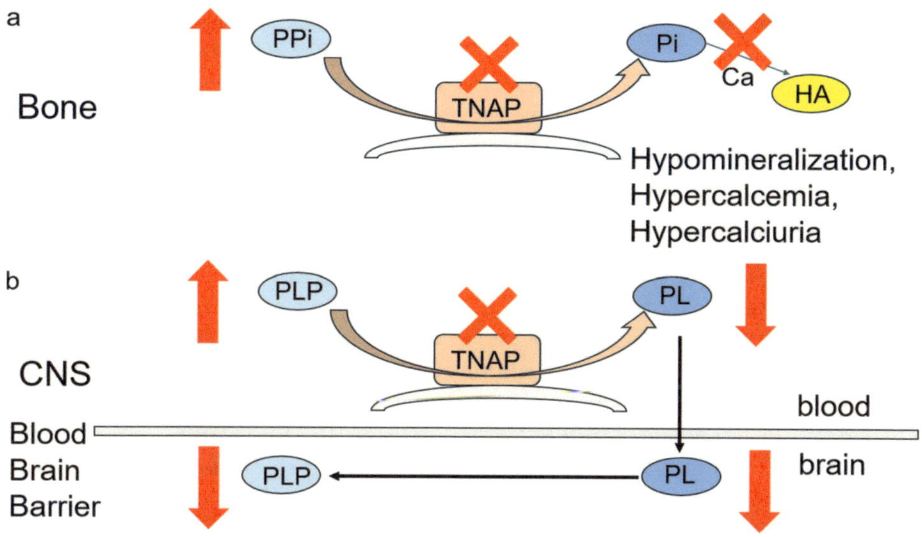

Figure 1.
Mechanisms of tissue nonspecific alkaline phosphatase (TNAP) in bone mineralization and brain function: In the bone, TNAP functions as a pyrophosphatase in bone tissue, producing phosphate (Pi) from pyrophosphate (PPi), which binds to calcium and deposits on hydroxyapatite, facilitating ossification (a). In the brain, TNAP functions as a pyrophosphatase in the blood converting pyridoxal 5′-phosphate (PLP) to pyridoxal (PL). Only PL crosses the blood–brain barrier, where it is reconverted to PLP to act as a neurotransmitter metabolism coenzyme (b).

a family linkage, remains a challenge [1, 5, 23]. Diminished TNAP activity leads to accumulation of TNAPs substrates including PPi, PLP, and PEA. Besides, TNAP is known as the liver-bone-kidney type ALP because it is abundantly expressed in those tissues. However, it is also extensively distributed throughout other organs. Other notable tissues that may also contribute to serum ALP activity through the expression of distinct ALP encoding genes (*ALPI, ALPP, ALPPL2*) include the small intestine, placenta, and germ cells.

In the body, PPi binds directly to hydroxyapatite crystals, thereby inhibiting ectopic calcification of soft tissue. In contrast, in the skeleton and teeth, TNAP hydrolyzes PPi to inorganic phosphate, thereby promoting crystal growth. The resulting phosphate binds to calcium to form hydroxyapatite; thus, the calcium that is not utilized for bone formation remains in the bloodstream, leading to hypercalcemia and hypercalciuria (**Figure 1a**).

Furthermore, PLP is involved in the synthesis of neurotransmitters in the brain. Because PLP cannot cross the blood–brain barrier, it is first dephosphorylated by TNAP to pyridoxal (PL). After crossing the blood brain barrier, PL is re-phosphorylated in the brain to PLP for utilization as an active cofactor of glutamate decarboxylases (**Figure 1b**). Thus, decreased TNAP activity is associated with reduced PLP levels in the brain, leading to vitamin B6-dependent convulsion [24–26].

3. Clinical presentation

Typically, HPP is classified according to the time of onset and severity, as presented below. However, as the disease progresses, the symptoms may increase, leading to a change in diagnosis [27]. Moreover, diagnosis is challenging because the phenotype does not always correlate with the type of genetic variant [5].

3.1 Perinatal hypophosphatasia

Typically, perinatal HPP [28] is fatal because of recurrent *in utero* fractures and respiratory complications arising from decreased thoracic circumference malformations, regardless of normal trunk length [29–31], as is shown in **Figure 2** [32]. However, there has been a paradigm shift with the advent of ERT, which has significantly improved survival rates and outcomes in patients with perinatal HPP. Perinatal HPP usually results from homozygous or compound heterozygous mutations of the *ALPL* gene, following an autosomal recessive inheritance pattern. On radiographs, it is evident that bone calcification manifests randomly [23]. For instance, some vertebrae may display calcification, whereas other vertebrates may not. In the skull, the frontal bone typically undergoes initial ossification; among the bones of the body, the clavicle represents the initial ossification site. Nevertheless, bones throughout the body are typically sufficiently thin to avoid detection on radiographs. Symmetrical defects in metaphysis of long bones such as femurs are characteristic of HPP, a condition referred to as central "tongue" [33].

Bowdler spurs represent a prevalent osseous characteristic of HPP [34–38]; they manifest as osteophytes projecting transversely from long tubular bones with deep skin dimples. However, the presence of Bowdler spurs alone does not warrant the exclusion of other diseases [35].

Osteogenesis imperfecta (OI) is more common [39] than HPP and should be differentiated based on the fetal bone abnormalities. OI manifests as a bone abnormality

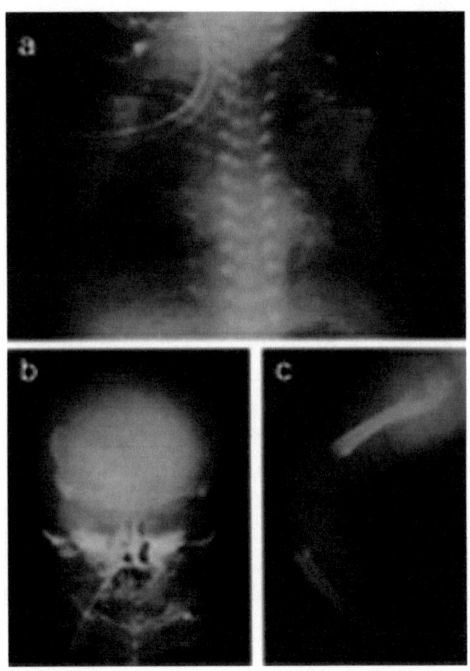

Figure 2.
Radiographs of a neonatal hypophosphatasia born at 42 weeks showing ossification failure and respiratory complications. Cesarean delivery was performed at 42 weeks, and 1-day gestation following fetal ultrasound showed extremity shortening. The baby received respiratory support from birth until death at 5 months due to respiratory failure. Clinical findings included vertebral ossification failure, and thoracic narrowing (a), skull ossification failure (b), and rickets-like changes in the femur (c). Reproduced with permission from Hirayama, T. et al.

during the fetal stage, and bisphosphonates are used to suppress bone resorption and improve ossification [40]. Conversely, HPP presents as a condition in which ossification is inherently impaired, rendering the application of bisphosphonates futile; thus, the use of bisphosphonates should be avoided [27]. Distinguishing these conditions is imperative, even in cases of prenatal bone anomalies.

3.2 Benign perinatal hypophosphatasia

Fetal echogenic bone abnormalities are detected prenatally, with a favorable prognosis characterized by improvement in bone abnormalities postnatally or in the third trimester of pregnancy [1, 5, 41]. Numerous studies on benign perinatal HPP [42] have been published in Japan, with some studies suggesting the effectiveness of treatment with asfotase alfa [43, 44]. However, treatment, particularly for rare diseases, is expensive and exerts pressure on the healthcare economy [45]. Ongoing discussions and research are crucially important to establish eligibility criteria for treatment among patients and explore potential alternatives for improved therapeutic outcomes [46].

3.3 Infantile hypophosphatasia

Infantile HPP [28] occurs within the first 6 months of life [1]. It usually results from homozygous or compound heterozygous mutations of the *ALPL* gene [27], following an autosomal recessive inheritance pattern. Craniosynostosis, severe skeletal abnormalities, hypercalcemia, hypercalciuria, and vitamin B6-dependent convulsion are the most common symptoms and are usually severe. Symptoms of infantile HPP are similar to those of perinatal HPP; however, the progression is comparatively slower, resulting in an extended lifespan for affected patients compared with those with perinatal HPP.

3.4 Childhood hypophosphatasia

HPP occurring after 6 months of age is typically classified as childhood HPP [47] and typically characterized by premature loss of fully developed teeth. It usually results from homozygous, compound heterozygous, or heterozygous mutations of the *ALPL* gene, following an autosomal recessive inheritance pattern. In addition, patients with childhood HPP frequently present with bone pain, muscle weakness, limb deformities, shortening, and dental symptoms. Furthermore, childhood HPP is characterized by tooth loss, primarily the front teeth, before age four, with the roots remaining intact [48, 49]. The median age at symptom onset in patients with autosomal recessive disease is 1 year, which is younger than that in patients with autosomal dominant disease (4 years). Blood levels of PPi and PLP were notably elevated in patients with recessively inherited conditions. However, clinical manifestations such as delayed ambulation and skeletal alterations were comparable between the two groups, although a higher incidence of bone fractures was observed among patients with the dominantly inherited disease [50]. Childhood HPP is nonfatal; however, it reduces quality of life and disrupts the ability to lead a typical childhood.

3.5 Adult-onset hypophosphatasia

Patients with adult-onset HPP [51] frequently complain of widespread chronic pain and muscle weakness [52, 53]; however, the pathophysiology remains unclear.

These patients may constitute undiagnosed cases of pediatric HPP; nonetheless, they often lack distinctive clinical features for identification. Symptoms, such as delayed fracture healing and those resembling osteoporosis, are commonly observed, along with pseudofractures and dysplasia of the bony bridges. Misdiagnosis as osteoporosis is frequent when HPP is not considered in the differential diagnosis [52, 54]. Bisphosphonates, one of the basic therapies for osteoporosis, is contraindicated in HPP [55, 56], making it crucial to include HPP in the differential diagnosis. Asfotase alfa has been shown to enhance bone mineralization and facilitate the healing of delayed union fractures in adult patients [57–61]. Moreover, ERT can restore muscle strength [62, 63]. Given the nonspecific symptoms observed in adult patients with HPP and the frequent misinterpretation of imaging results resembling osteoporosis, assessment of serum ALP levels is a valuable initial diagnostic approach [54]. In addition, a single diagnostic test is often inadequate. Notably, a substantial proportion of patients with milder forms of HPP show improvement with supportive treatments alone [64, 65]. However, a subset of patients requires ERT. Judicious selection of patients eligible for ERT poses an ongoing challenge in clinical management [66].

3.6 Odontohypophosphatasia

In odontohypophosphatasia, a notable feature is premature tooth loss at a young age, despite the absence of bone abnormalities [7]. The root of the fallen tooth remains intact, and pathology reveals cementum dysplasia. Typically, poor calcification of both the acellular cementum and dentin is observed [48]. However, it is challenging to prevent the losing of primary teeth upon initiation of ERT, which may commence within days of birth in cases of timely diagnosis. This difficulty arises because the primary teeth are formed during fetal development [67]. Therefore, absent teeth cannot be transplanted, necessitating denture placement. Although the mild variant of the condition has a more favorable prognosis than its severe counterpart, it nonetheless represents a significant detriment to the patient's quality of life. Cases initially diagnosed solely based on dental symptoms or considered odontohypophosphatasia may later be re-diagnosed as pediatric or adult-onset HPP after presentation of bone or muscle symptoms [49]. Studies indicate that ERT has positive effects on dental symptoms associated with infantile and childhood HPP [68, 69].

4. Treatment

Before the advent of ERT, only symptomatic management options were available. For severe perinatal HPP, treatment options were limited to ventilation and low-calcium milk for hypercalcemia, hypercalciuria, and vitamin B6-dependent convulsions [1]. Numerous modalities have been explored to address the needs of individuals with HPP. These interventions include bone marrow cell transplantation [70, 71], transplantation of bone fragments along with cultured osteoblasts [72], administration of teriparatide (recombinant human parathyroid hormone PTH 1–34) [73, 74], ERT utilizing ALP-enriched serum derived from patients with Paget's disease of the bone [75], and infusion of plasma sourced from healthy donors [76] or ALP purified from hepatic tissue [77]. Despite these interventions, noticeable clinical and radiographic enhancements have been observed only in some patients, reflecting the limited efficacy of these therapeutic modalities.

The clinical manifestations observed in TNAP knockout (*Alpl*$^{-/-}$) mice closely mimic those seen severe infantile HPP, making *Alpl*$^{-/-}$ mice an optimal animal model for studying HPP [78]. Although initially appearing phenotypically normal at birth, these mice progress to exhibit growth failure, vitamin B6-dependent convulsions, and hypomineralization, typically dying before weaning. *Alpl*$^{-/-}$ mice were rescued by daily subcutaneous administration of a bone-targeted variant of TNAP obtained by conjugating a bone-targeting deca-aspartate sequence to the C-terminus of soluble TNAP (TNAP-D_{10}) [79]. These promising preclinical outcomes paved the way for the clinical implementation of ERT in HPP patients [80]. This significant advancement has not only extended survival rates in patients with severe infantile HPP but also enhanced the quality of life for patients with pediatric and adult-onset HPP [81].

ERT significantly enhances prognosis, particularly for patients with severe perinatal HPP. However, as the enzyme degrades over time, continuous replenishment by subcutaneous injections 3–6 times a week is required [82]. Patients seek to alleviate the persistent pain due to the frequent injections and receive definitive curative therapy [83].

Furthermore, the use of ERT is associated with some challenges, such as the development of lipodystrophy at the site of injections and the development of antidrug antibodies [84]. Improvements in infantile HPP cases following bone marrow transplantation have been reported. In addition, allogeneic mesenchymal stem cell transplantation has demonstrated efficacy in fatal perinatal HPP [85]. However, the treatment is accompanied with pre-therapy procedures, such as chemotherapy, radiation therapy, and immunotherapy, and associated with potential posttransplant complications, such as graft-versus-host disease.

5. Gene therapy

5.1 History

Gene therapy was initially proposed as a radical treatment for diseases *via* repair of damaged genes. However, this modality proved challenging. In the early days of gene therapy, recognizing target genes was challenging, and gene modification techniques were insufficiently developed, making it difficult to alter target genes [86, 87]. Alternative approaches to deliver functionally missing molecules were necessary. Not all genetic diseases can be treated using gene therapy. However, gene therapy can serve as a potentially curative approach for diseases that meet two criteria: 1) the condition must be attributable to a single gene and 2) the expression of that gene can reach a therapeutic level.

The direct introduction of a functional gene into the body is known as *in vivo* therapy. The vectors employed for gene delivery include viral and nonviral vectors such as plasmid vectors, bacterial vectors, and lipoplexes. At present, the primary clinical application of viral vectors is therapeutic because of their high transduction efficiency. Although the use of viral vectors is simple and efficient, it requires a large viral dose and there is a possibility that the viral vectors may enter organs besides the target organs. Meanwhile, *ex vivo* therapy describes the process of cell collection, culturing, and transduction of functional genes prior to their administration to the patient. *Ex vivo* therapy is applicable in certain primary immunodeficiency diseases, congenital errors in metabolism, and other conditions that necessitate hematopoietic stem cell transplantation [88, 89]. CAR-T cell therapy [90], which exhibits antitumor effects against refractory B-cell tumors, has been classified as an *ex vivo* therapy. It is expected to reduce side effects, such as secondary malignancy after CAR-T cell therapy, by allowing therapeutic cells to grow outside the body before returning to them and avoiding the transcription start site, thereby making gene therapy safer. Patient burden is substantial because of the necessity of pre-treatment before hematopoietic stem cell gene therapy.

In the years leading up to the 1990s, various attempts were made to use gene therapy for genetic diseases and cancer with the goal of achieving curative treatment. These attempts, however, had limited success. The first notable success was achieved in 2000 [91, 92]. Nevertheless, safety concerns arose a few years later because of incidences of leukemia development [93] and deaths resulting from other treatments [94, 95], thereby dampening public expectations. As a result of these limitations, advancements in techniques characterized by enhanced safety measures and ethical considerations have resulted in documented success since the late 2000s. For example, Zolgensma®, an *in vivo* gene therapy utilizing an adeno-associated virus (AAV) vector, significantly enhanced life expectancy and motor function in patients with spinal muscular atrophy, a condition previously lacking curative treatment. Although this represents a significant advancement in healthcare, the associated substantial expenses remain a significant challenge.

Studies have focused on genome-editing technologies, such as CRISPR-Cas9, to correct genetic defects caused by gain-of-function mutations in dominant genetic disorders [96]. CRISPR-Cas9 is a technology used for cutting and editing specific sections of DNA and has been clinically applied to treat sickle cell disease and beta-thalassemia. This method precisely targets and modifies faulty genes, thereby offering potential solutions for previously untreatable conditions. Casgevy® for beta-thalassemia [97, 98] was approved in Europe, the UK, and the USA. Extensive research and clinical trials are necessary to confirm the safety and efficacy of these techniques before wide application in clinical practice. In addition, CAR-T therapy involves genetically modifying a patient's T cells to recognize and attack specific cancer cells, primarily in blood cancers, such as multiple myeloma and lymphoma [99]. Kimliah® (CD19-CAR), Abecma® (BCMA-CAR), and nine other CAR-T therapy drugs have been approved. This therapy has been primarily applied in clinical settings for the treatment of blood cancer. Furthermore, oncolytic virus therapy involves the introduction of a natural or genetically engineered virus into the uterus, where it selectively replicates in the tumor tissue and targets tumor cells for destruction [100].

5.2 Challenges with gene therapy

Gene therapy faces several challenges, particularly regarding the use of different vectors. Local administration of AAV vectors is relatively safe, as seen with interventions like Luxturna® and Upstaza®, which require smaller doses and prevent significant immune responses. Luxturna® (1.5×10^{11} vg/eye) is an intraretinal injection of voretigene neparvovec-rzyl [101] gene therapy drug, and Upstaza® (1.81×10^{11} vg total) is an intraputamen injection of an AAV vector containing the human *AADC* gene (AAV2-hAADC) [102] gene therapy drug. However, systemic administration presents greater risks, including the requirement for large quantities of vectors. In cases of X-linked myotubular myopathy, the systemic administration of AAV8 at high doses results in fatal liver failure in some patients. At a low dose of 1.3×10^{14} vg/kg in seven patients and at a high dose of 3.5×10^{14} vg/kg in 17 patients, three patients from the high-dose group and one from the low-dose group experienced fatal liver failure [103–106]. Similarly, in the treatment of Duchenne muscular dystrophy, high doses of intravenous recombinant AAV (rAAV) cause fatal immune responses and capillary leakage [107, 108]. The long-term sustainability and safety of AAV vectors remain uncertain, with the risks of inappropriate gene expression and immune responses. In addition, treatment efficacy relies on the absence of antibodies in the body that can neutralize AAV vectors (negative neutralizing antibodies), thus preventing treatment with AAV vectors [109].

Although advancements in gene therapy are underway, researchers must be mindful of uncharted areas that require exploration. To date, no study has established an association between systemic administration of AAV vectors and gene integration into germ cells. However, owing to the inability to examine all germ cells, a thorough confirmation of the absence of germ cell modification is essential [110, 111].

Lentiviral vectors, which integrate genes into random chromosomal locations, pose a risk of insertional mutagenesis, potentially leading to leukemia [112]. However, advancements in research have made self-inactivating (SIN) lentiviral vectors safer, eliminating the occurrence of leukemia. Recent clinical trials using SIN lentiviral vectors have brought them closer to clinical application [112, 113]. Other viral vectors can induce immune responses that diminish the efficacy of gene therapy and cause adverse effects. Although nonviral vectors pose fewer safety concerns and immune responses, they are associated with issues pertaining to efficiency and targeted gene delivery to specific tissues.

The high costs of gene therapy, particularly for rare diseases, pose a significant barrier. For example, Glybera®, a gene therapy for lipoprotein lipase deficiency approved in 2012, was discontinued in 2017 owing to low cost-effectiveness despite its high price [114]. The allocation of medical expenses for rare diseases is, therefore, heavily influenced by societal values and political policies. Gene therapy drugs such as LUXTURNA® for Leber congenital amaurosis, Zolgensma® for spinal muscular atrophy (SMA), and Hemgenix® for hemophilia B are notably expensive, costing $850,000 for both eyes, $2.1 million per injection, and over $3.5 million per injection, respectively. The high development costs coupled with the limited patient population for rare diseases contribute to the high prices of these treatments. To address these issues, preclinical trials in the United States are exploring the potential of modifying genes within the same viral vector to treat multiple diseases [115, 116]. Collaborative efforts between public and private sectors are crucial to advancing clinical development and enhancing efficiency.

5.3 HPP gene therapy

Despite the availability of ERT as an effective treatment for HPP, the discomfort associated with frequent injections and injection site reactions pose challenges for patients. Therefore, we aimed to employ gene therapy techniques to enable the patient's body to continuously and autonomously produce enzymes.

Intravenous therapy using AAV8 viral [117] and lentiviral vectors [118], fetal therapy using AAV9 viral vectors [119], bone marrow transplantation using lentiviral vectors [120], and intramuscular therapy using AAV8 viral vectors [121] have all demonstrated success in an HPP model, $Alpl^{-/-}$ mice. Bone marrow transplantation is challenging because of the risk of leukemia and burden of pretreatment on patients. Systemic administration of AAV is avoided to prevent systemic immune reactions and antibody production. Consequently, intramuscular injection is selected as the preferred method. AAV vectors offer safe and efficient gene delivery to nondividing cells such as muscle cells, liver cells, and neurons, though the size limitation of inserted genes and the challenge of efficient production [103]. Moreover, AAV vectors exhibit tropism, allowing for the use of different subtypes tailored to specific organs [122]. Additionally, employing a tissue-specific promoter ensures that gene expression occurs exclusively within the targeted organ [122, 123]. Based on our previous study, intramuscular injection therapy using AAV8 viral vectors has emerged as the safest, most practical, and simplest approach.

5.4 Toward clinical application: AAV8-TANP-D10

Leveraging AAV organ tropism, we aimed to utilize AAV8, known for its tropism to infect liver and muscle tissues [122, 124], for localized injection into the muscle. The advantages of local administration include simplicity and safety, low risk of migration to other organs, suppression of immune responses associated with intravenous administration (i.e., antibody production and reactions that affect the whole body), and low potential for tumor formation. Furthermore, if adverse events cause the patient to discontinue enzyme expression, the muscle at the inoculation site can be surgically removed in the worst-case scenario.

AAV8-TNAP-D_{10}, a viral vector product expressing TNAP, is a gene therapy designed to extend lifespan, suppress symptom progression, and ameliorate symptoms in patients with HPP *via* a single intramuscular injection. AAV8-TNAP-D_{10} is an AAV8-TNAP vector with deca-aspartate (D_{10}) that targets the bone using a non-tissue-specific CAG promoter [121]. A single injection of AAV8-TNAP-D_{10} was administered into the thigh muscle of neonatal *Alpl*$^{-/-}$ mice (**Figure 3a**). Administration of AAV8-TNAP-D_{10} led to increased weight and a healthier appearance in *Alpl*$^{-/-}$ mice compared to untreated *Alpl*$^{-/-}$ mice (**Figure 3b**). The injection of AAV8-TNAP-D_{10} into the thigh muscle of neonatal *Alpl*$^{-/-}$ mice extended their lifespan and sustained elevated ALP activity in moribund *Alpl*$^{-/-}$ mice for 18 months at a dose of $\geq 3.0 \times 10^{11}$ vg/body. Micro-CT examination showed that bone maturation in treated *Alpl*$^{-/-}$ mice was comparable to that of wild-type mice (**Figure 3c**). Micro-CT examination of *Alpl*$^{-/-}$ mice's mandibles revealed that the mandibular structure in treated mice improved to the level of the wild-type mice. Moreover, there were notable improvements in the cementum and periodontal ligament structures compared to those in untreated mice. Enamel thickness, pulp structure volume, and thickness improved posttreatment to a degree where they could be assessed as comparable to those of the wild-type mice [125].

The use of *Alpl* $^{Prx1/Prx1}$ mice allows long-term studies of bones and teeth that are not possible with untreated lethal *Alpl*$^{-/-}$ mice [126]; moreover, they can be used as an adult HPP model or for tooth analysis. Treatment of *Alpl* $^{Prx1/Prx1}$ mice with AAV8-TNAP-D_{10} improves ossification, thus demonstrating that AAV8-TNAP-D_{10} is effective even in adult HPP models [127].

Prior to clinical trials, efficacy must be demonstrated in rodents and larger animals. A sheep model of HPP has been established in large animals using CRISPR-Cas9 technology to introduce a missense mutation (c.1077C > G; p.I359M), which serves as an effective model for studying alveolar bone conditions [128], but gene therapy has not yet been explored in that model. We selected the cynomolgus monkey (Macaca fascicularis) as a large animal model to assess the effectiveness of AAV8-TNAP-D_{10}. Our objectives were to 1) evaluate whether a single intramuscular injection of AAV8-TNAP-D_{10} maintained plasma ALP activity, 2) assess any potential antibody-mediated rejection of the protein expressed in animals with differing immune systems, and

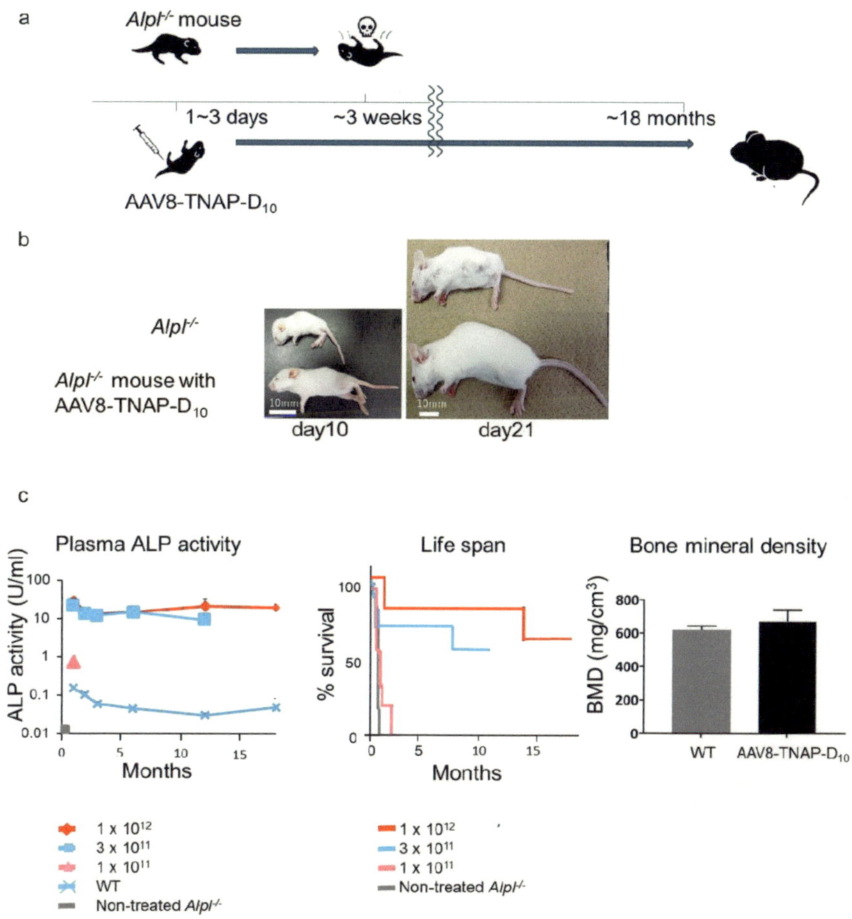

Figure 3.
Impact of one-time intramuscular AAV8-TNAP-D_{10} treatment on survival in Alp$^{-/-}$ mice. Alpl$^{-/-}$ mice injected with a single intramuscular injection of AAV8-TNAP-D_{10} in the quadriceps muscle within 3 days after birth ($1.0x10^{12}$ vg/body; n = 7, $3.0x10^{11}$ vg/body; n = 7, $1.0x10^{11}$ vg/body; n = 5) were observed for 18 months (a). Untreated model mice typically died before weaning. Physical appearance of untreated Alpl$^{-/-}$ mice vs. treated Alpl$^{-/-}$ mice (b). Plasma ALP activity, the survival curves, and bone mineral density at 18 months with computer tomography reconstruction of the femur bone. Bone mineral density at 18 months of age. CT reconstruction of the femur in AAV8-TNAP-D_{10} treated Alpl$^{-/-}$ mice ($1.0x10^{12}$ vg/body) vs. wild type (c).

3) examine adverse effects such as ectopic calcification, organ damage, or carcinogenesis, which have been previously reported following gene therapy failures.

AAV8-TNAP-D_{10} was injected into the lateral vastus lateralis muscle of two macaque monkeys under anesthesia. One monkey received 1×10^{13} vg/body and was observed for 266 days, whereas the other received 4×10^{13} vg/body (1×10^{13} vg at four sites) and was observed for 196 days. Both monkeys were compared to the controls. The rationale for increasing the number of the administration site instead of solely increasing the dosage was prevention of the leakage of the injected substance into the bloodstream from the muscle site, which could lead to inadvertent intravenous administration. The results showed that monkeys had higher serum ALP activity than the controls, which was maintained throughout the observation period. Although anti-TNAP antibodies appeared and caused a temporary decrease in serum ALP activity, these antibodies eventually decreased, thereby maintaining high ALP activity. Anti-AAV antibodies also appeared but did not affect ALP activity. Regarding side effects, all blood tests, including those for liver function (aspartate aminotransferase, alanine aminotransferase), renal function (blood urea nitrogen, creatinine), and calcium metabolism (calcium, phosphate), showed no abnormalities. There were no pathological findings (no tumor formation or ectopic calcification), either macroscopically or microscopically. No local inflammation was observed at the injection site. Vector biodistribution was confirmed in each organ; furthermore, AAV8-TNAP-D_{10} was detected only in the skin and muscles, where it was administered (ongoing research and unpublished data).

6. Conclusions

We have discussed the treatment of HPP, the history of gene therapy, and our review of gene therapy for HPP. Although the approval of ERT has marked a revolutionary advancement, its reliance on regular and frequent injections significantly impacts the quality of life of patients and their families. Therefore, alternative treatments are urgently required.

In our study, AAV8-TNAP-D_{10} administered *via* a single intramuscular injection in both the early- and later-onset HPP mouse models demonstrated sustained elevation of serum ALP activity throughout the lifespan, along with improvements in bone and tooth ossification. No adverse effects were observed within this treatment range. In addition, AAV8-TNAP-D_{10} was shown to be safe and effective in primates. Thus, AAV8-TNAP-D_{10} intramuscular muscle injection treatment appears to have promising efficacy and safety profile for clinical application.

Given the costs associated with gene therapy as well as the broad spectrum of symptoms in patients with HPP, it is crucial to address societal considerations regarding patient selection for gene therapy.

Our future studies will include safety assessments and further investigations in preparation for clinical trials. We intend to advance our research efforts in the pursuit of enhanced safety profiles.

Acknowledgements

We sincerely thank Dr. Dongway Zhao and Dr. Noriko Miyake for their technical assistance and helpful discussions that greatly contributed to this work. This research was supported by AMED (Japan Agency for Medical Research and Development) under Grant Number JP23ek0109670 to KM and NIDCR (National Institute of Dental and Craniofacial Research) of the NIH (National Institutes of Health) under Grant Number DE12889 to JLM. This work was also supported by JSPS (Japan Society for the Promotion Science) KAKENHI Grant Numbers JP 20 K08268 and JP 15 K09605 to TM and KM.

Conflict of interest

TM and KM have previously received research funding from Aruvant Sciences and held an endowed chair affiliated with Aruvant Sciences during the course of the research. JLM received partial research funding from Aruvant Sciences.

DOI: http://dx.doi.org/10.5772/intechopen.1006268

Author details

Tae Matsumoto[1,2]*, Sonoko Narisawa[3], José Luis Millán[3] and Koichi Miyake[2]

1 Department of Pediatrics, Nippon Medical School, Tokyo, Japan

2 Department of Gene Therapy, Nippon Medical School, Tokyo, Japan

3 Sanford Children's Health Research Center, Sanford Burnham Prebys Medical Discovery Institute, La Jolla, CA, United States

*Address all correspondence to: tae@nms.ac.jp

References

[1] Mornet E. Hypophosphatasia. Orphanet Journal of Rare Diseases. 2007;**2**:40

[2] Hypophosphatasia ME. Best Practice & Research. Clinical Rheumatology. 2008;**22**(1):113-127

[3] Millán JI. The in Vivo Role of TNAP. Mammalian Alkaline Phosphatases: From Biology to Applications in Medicine and Biotechnology. Weinheim: Federal Republik of Germany: WILEY-VCH Verlag GmbH & Co.KGaA; 2006. pp. 105-129

[4] Hypophosphatasia FD. The American Journal of Medicine. 1957;**22**(5):730-746

[5] Mornet E. Genetics of hypophosphatasia. Archives de Pédiatrie. 2017;**24**(5S2):5S51-55S6

[6] Taketani T, Onigata K, Kobayashi H, Mushimoto Y, Fukuda S, Yamaguchi S. Clinical and genetic aspects of hypophosphatasia in Japanese patients. Archives of Disease in Childhood. 2014;**99**(3):211-215

[7] Reis FS, Lazaretti-Castro M. Hypophosphatasia: From birth to adulthood. Arch Endocrinol Metab. 2023;**67**(5):e000626

[8] Watanabe A, Karasugi T, Sawai H, Naing BT, Ikegawa S, Orimo H, et al. Prevalence of c.1559delT in ALPL, a common mutation resulting in the perinatal (lethal) form of hypophosphatasia in Japanese and effects of the mutation on heterozygous carriers. Journal of Human Genetics. 2011;**56**(2):166-168

[9] Mornet E, Yvard A, Taillandier A, Fauvert D, Simon-Bouy B. A molecular-based estimation of the prevalence of hypophosphatasia in the European population. Annals of Human Genetics. 2011;**75**(3):439-445

[10] Briot K, Roux C. Adult hypophosphatasia. Archives de Pédiatrie. 2017;**24**(5S2):5S71-55S3

[11] Inoue D. Diagnosis and treatment of adult hypophosphatasia: Still a big challenge? Osteoporos Sarcopenia. 2024;**10**(1):1-2

[12] Kim SM, Korkmaz F, Sims S, Ryu V, Yuen T, Zaidi M. Musculoskeletal and neurocognitive clinical significance of adult hypophosphatasia. Osteoporos Sarcopenia. 2023;**9**(4):115-120

[13] Inci A, Ergin FBC, Yuce BT, Ciftci B, Demir E, Buyan N, et al. Hypophosphatasia: Is it an underdiagnosed disease even by expert physicians? Journal of Bone and Mineral Metabolism. 2021;**39**(4):598-605

[14] Farman MR, Rehder C, Malli T, Rockman-Greenberg C, Dahir K, Martos-Moreno GA, et al. The global ALPL gene variant classification project: Dedicated to deciphering variants. Bone. 2024;**178**:116947

[15] The ALPL Gene Variant Database [Internet]. [Cited 2024 May 20]. Available from: https://alplmutationdatabase.jku.at/

[16] Asfotase Alfa (Strensiq). Common Drug Review. Ottawa: Canadian Agency for Drugs and Technologies in Health; 2017

[17] Naslonski T. Laiken, Rowan and their Family's HPP Journey 2020 [Updated November 6, 2020 Cited 2024 June 13].

Available from: https://softbones.org/laiken-rowan-and-their-familys-hpp-journey/

[18] Castells Vilella L, Sanchez-Pintos P, Muniz Llama JF, Gamez Martinez M, Couce ML, Anton J. Age- and sex-dynamic fluctuations and reference intervals for alkaline phosphatase among the Spanish population. Archives of Pathology & Laboratory Medicine. 2024 [Online ahead of print]

[19] Garcia-Fontana C, Villa-Suarez JM, Andujar-Vera F, Gonzalez-Salvatierra S, Martinez-Navajas G, Real PJ, et al. Epidemiological, clinical and genetic study of Hypophosphatasia in a Spanish population: Identification of two novel mutations in the Alpl gene. Scientific Reports. 2019;**9**(1):9569

[20] Joseph JC, Baker C, Sprang ML, Bermes EW. Changes in plasma proteins during pregnancy. Annals of Clinical and Laboratory Science. 1978;**8**(2):130-141

[21] Takahashi Y, Sawai H, Murotsuki J, Satoh S, Yamada T, Hayakawa H, et al. Parental serum alkaline phosphatase activity as an auxiliary tool for prenatal diagnosis of hypophosphatasia. Prenatal Diagnosis. 2017;**37**(5):491-496

[22] Medicine NLo. ALPL Alkaline Phosphatase, Biomineralization Associated [Homo Sapiens (Human)] 2024 May 17 [Cited 2024 June 22]. Available from: https://www.ncbi.nlm.nih.gov/gene/249

[23] Ikenoue S, Miyakoshi K, Ishii T, Sato Y, Otani T, Akiba Y, et al. Discordant fetal phenotype of hypophosphatasia in two siblings. American Journal of Medical Genetics. Part A. 2018;**176**(1):171-174

[24] Fonta C, Negyessy L, Renaud L, Barone P. Areal and subcellular localization of the ubiquitous alkaline phosphatase in the primate cerebral cortex: Evidence for a role in neurotransmission. Cerebral Cortex. 2004;**14**(6):595-609

[25] Waymire KG, Mahuren JD, Jaje JM, Guilarte TR, Coburn SP, MacGregor GR. Mice lacking tissue non-specific alkaline phosphatase die from seizures due to defective metabolism of vitamin B-6. Nature Genetics. 1995;**11**(1):45-51

[26] Sebastian-Serrano A, Engel T, de Diego-Garcia L, Olivos-Ore LA, Arribas-Blazquez M, Martinez-Frailes C, et al. Neurodevelopmental alterations and seizures developed by mouse model of infantile hypophosphatasia are associated with purinergic signalling deregulation. Human Molecular Genetics. 2016;**25**(19):4143-4156

[27] Mornet E, Taillandier A, Domingues C, Dufour A, Benaloun E, Lavaud N, et al. Hypophosphatasia: A genetic-based nosology and new insights in genotype-phenotype correlation. European Journal of Human Genetics. 2021;**29**(2):289-299

[28] Vernon HJ. Hypophosphatasia, Infantile; HPPI 2021 [Updated 10/19/2021; Cited 2024 July 15]. Available from: https://omim.org/entry/241500

[29] Taketani T, Kanai R, Abe M, Mishima S, Tadokoro M, Katsube Y, et al. Therapy-related Ph+ leukemia after both bone marrow and mesenchymal stem cell transplantation for hypophosphatasia. Pediatrics International. 2013;**55**(3):e52-e55

[30] Leung EC, Mhanni AA, Reed M, Whyte MP, Landy H, Greenberg CR. Outcome of perinatal hypophosphatasia in Manitoba mennonites: A retrospective cohort analysis. JIMD Reports. 2013;**11**:73-78

[31] Martos-Moreno GA, Rockman-Greenberg C, Ozono K, Petryk A, Kishnani PS, Dahir KM, et al. Clinical profiles of children with Hypophosphatasia prior to treatment with enzyme replacement therapy: An observational analysis from the global HPP registry. Hormone Research in Pædiatrics. 2023;**97**(3):233-242

[32] Hirayama T, Kuwabara K, Tsugu H, Hirayama T, 平山恒徳, 桑原健太郎, et al. Infantile hypophosphatasia 乳児型低ホスファターゼ症. Pediatrics of Japan 小児科. 1993;**34**(12):the front matter

[33] Offiah AC, Vockley J, Munns CF, Murotsuki J. Differential diagnosis of perinatal hypophosphatasia: Radiologic perspectives. Pediatric Radiology. 2019;**49**(1):3-22

[34] Oestreich AE, Bofinger MK. Prominent transverse (Bowdler) bone spurs as a diagnostic clue in a case of neonatal hypophosphatasia without metaphyseal irregularity. Pediatric Radiology. 1989;**19**(5):341-342

[35] Oestreich AE. Bowdler spur also found in camptomelic dysplasia. Pediatric Radiology. 2016;**46**(2):300

[36] Kozlowski K, Sutcliffe J, Barylak A, Harrington G, Kemperdick H, Nolte K, et al. Review of 24 cases. Pediatric Radiology. 1976;**5**(2):103-117

[37] Wenkert D, McAlister WH, Coburn SP, Zerega JA, Ryan LM, Ericson KL, et al. Hypophosphatasia: Nonlethal disease despite skeletal presentation in utero (17 new cases and literature review). Journal of Bone and Mineral Research. 2011;**26**(10):2389-2398

[38] Guguloth A, Aswani Y, Anandpara KM. Prenatal diagnosis of hypophosphatasia congenita using ultrasonography. Ultrasonography. 2016;**35**(1):83-86

[39] Marom R, Rabenhorst BM, Morello R. Osteogenesis imperfecta: An update on clinical features and therapies. European Journal of Endocrinology. 2020;**183**(4):R95-R106

[40] Botor M, Fus-Kujawa A, Uroczynska M, Stepien KL, Galicka A, Gawron K, et al. Osteogenesis Imperfecta: Current and prospective therapies. Biomolecules. 2021;**11**(10):1493

[41] Mornet E. Molecular genetics of Hypophosphatasia and phenotype-genotype correlations. Sub-Cellular Biochemistry. 2015;**76**:25-43

[42] Bangura A, Wright L, Shuler T. Hypophosphatasia: Current literature for pathophysiology, clinical manifestations, diagnosis, and treatment. Cureus. 2020;**12**(6):e8594

[43] Matsushita M, Mishima K, Nagata T, Kamiya Y, Imagama S, Kitoh H. Asfotase alfa has a limited effect in improving the bowed limbs in perinatal benign hypophosphatasia: A case report. Clinical Pediatric Endocrinology. 2021;**30**(1):53-56

[44] Kato H, Hidaka N, Koga M, Kinoshita Y, Makita N, Nangaku M, et al. Radiological evaluation of pseudofracture after the administration of asfotase alfa in an adult with benign prenatal hypophosphatasia: A case report. Bone Reports. 2022;**16**:101163

[45] Mavilio F. Developing gene and cell therapies for rare diseases: An opportunity for synergy between academia and industry. Gene Therapy. 2017;**24**(9):590-592

[46] Rush ET. Childhood hypophosphatasia: To treat or not to treat.

Orphanet Journal of Rare Diseases. 2018;**13**(1):116

[47] Vernon HJ. Hypophosphatasia, Childhood; HPPC 2021 [Cited 2024 July 15]. Available from: https://omim.org/entry/241510

[48] Okawa R, Nakano K. Dental manifestation and management of hypophosphatasia. Japanese Dental Science Review. 2022;**58**:208-216

[49] Okawa R, Kadota T, Matayoshi S, Nakano K. Dental manifestations leading to the diagnosis of Hypophosphatasia in two children. Journal of Dentistry for Children (Chicago, Ill.). 2020;**87**(3):179-183

[50] Kishnani PS, Del Angel G, Zhou S, Rush ET. Investigation of ALPL variant states and clinical outcomes: An analysis of adults and adolescents with hypophosphatasia treated with asfotase alfa. Molecular Genetics and Metabolism. 2021;**133**(1):113-121

[51] Vernon HJ. Hypophosphatasia, Adult; HPPA 2021 [Updated 06/04/2021; Cited 2024 July 15]. Available from: https://omim.org/entry/146300

[52] Briot K, Roux C. Adult hypophosphatasia. Current Opinion in Rheumatology. 2016;**28**(4):448-451

[53] Berkseth KE, Tebben PJ, Drake MT, Hefferan TE, Jewison DE, Wermers RA. Clinical spectrum of hypophosphatasia diagnosed in adults. Bone. 2013;**54**(1):21-27

[54] Ng E, Ashkar C, Seeman E, Schneider HG, Nguyen H, Ebeling PR, et al. A low serum alkaline phosphatase may signal hypophosphatasia in osteoporosis clinic patients. Osteoporosis International. 2023;**34**(2):327-337

[55] Sutton RA, Mumm S, Coburn SP, Ericson KL, Whyte MP. "Atypical femoral fractures" during bisphosphonate exposure in adult hypophosphatasia. Journal of Bone and Mineral Research. 2012;**27**(5):987-994

[56] Whyte MP. Atypical femoral fractures, bisphosphonates, and adult hypophosphatasia. Journal of Bone and Mineral Research. 2009;**24**(6):1132-1134

[57] Hidaka N, Murata H, Tachikawa K, Osaki K, Sekiyama T, Kinoshita Y, et al. The effect of Asfotase Alfa on plasma and urine pyrophosphate levels and Pseudofractures in a patient with adult-onset Hypophosphatasia. JBMR Plus. 2023;**7**(12):e10842

[58] Seefried L, Kishnani PS, Moseley S, Denker AE, Watsky E, Whyte MP, et al. Pharmacodynamics of asfotase alfa in adults with pediatric-onset hypophosphatasia. Bone. 2021;**142**:115664

[59] Sturznickel J, Schmidt FN, von Vopelius E, Delsmann MM, Schmidt C, Jandl NM, et al. Bone healing and reactivation of remodeling under asfotase alfa therapy in adult patients with pediatric-onset hypophosphatasia. Bone. 2021;**143**:115794

[60] Magdaleno AL, Singh S, Venkataraman S, Perilli GA, Lee YY. Adult-onset Hypophosphatasia: Before and after treatment with Asfotase Alfa. AACE Clinical Case Reports. 2019;**5**(6):e344-e3e8

[61] Klidaras P, Severt J, Aggers D, Payne J, Miller PD, Ing SW. Fracture healing in two adult patients with Hypophosphatasia after Asfotase Alfa therapy. JBMR Plus. 2018;**2**(5):304-307

[62] Koyama H, Yasuda S, Kakoi S, Ohata Y, Shimizu Y, Hasegawa C, et al.

Effect of Asfotase Alfa on muscle weakness in a Japanese adult patient of Hypophosphatasia with low ALP levels. Internal Medicine. 2020;**59**(6):811-815

[63] Rolvien T, Schmidt T, Schmidt FN, von Kroge S, Busse B, Amling M, et al. Recovery of bone mineralization and quality during asfotase alfa treatment in an adult patient with infantile-onset hypophosphatasia. Bone. 2019;**127**:67-74

[64] Mizuno H, Sawa N, Sekine A, Inoue N, Oba Y, Ikuma D, et al. A bone Histomorphometric analysis of Hypophosphatasia-related osteoporosis after Teriparatide treatment. Internal Medicine. 2023;**62**(1):75-79

[65] Righetti M, Wach J, Desmarchelier R, Coury F. Teriparatide treatment in an adult patient with hypophosphatasia exposed to bisphosphonate and revealed by bilateral atypical fractures. Joint, Bone, Spine. 2018;**85**(3):365-367

[66] Brandi ML, Khan AA, Rush ET, Ali DS, Al-Alwani H, Almonaei K, et al. The challenge of hypophosphatasia diagnosis in adults: Results from the HPP international working group literature surveillance. Osteoporosis International. 2024;**35**(3):439-449

[67] Lalys L, Ruquet M, Tardivo D, Laibi S, Bartoli C, Adalian P, et al. Estimation of gestational age from tooth germs: Biometric study of DentaScan images. Journal of Forensic Sciences. 2011;**56**(1):220-223

[68] Okawa R, Kokomoto K, Nakano K. Dental effects of enzyme replacement therapy in case of childhood-type hypophosphatasia. BMC Oral Health. 2021;**21**(1):323

[69] Okawa R, Matayoshi S, Kariya R, Ogaya Y, Nomura R, Nakano K. Effects of enzyme replacement therapy for primary teeth in a patient with infantile Hypophosphatasia. The Journal of Clinical Pediatric Dentistry. 2020;**44**(5):348-351

[70] Tadokoro M, Kanai R, Taketani T, Uchio Y, Yamaguchi S, Ohgushi H. New bone formation by allogeneic mesenchymal stem cell transplantation in a patient with perinatal hypophosphatasia. The Journal of Pediatrics. 2009;**154**(6):924-930

[71] Whyte MP, Kurtzberg J, McAlister WH, Mumm S, Podgornik MN, Coburn SP, et al. Marrow cell transplantation for infantile hypophosphatasia. Journal of Bone and Mineral Research. 2003;**18**(4):624-636

[72] Cahill RA, Wenkert D, Perlman SA, Steele A, Coburn SP, McAlister WH, et al. Infantile hypophosphatasia: Transplantation therapy trial using bone fragments and cultured osteoblasts. The Journal of Clinical Endocrinology and Metabolism. 2007;**92**(8):2923-2930

[73] Whyte MP, Mumm S, Deal C. Adult hypophosphatasia treated with teriparatide. The Journal of Clinical Endocrinology and Metabolism. 2007;**92**(4):1203-1208

[74] Camacho PM, Painter S, Kadanoff R. Treatment of adult hypophosphatasia with teriparatide. Endocrine Practice. 2008;**14**(2):204-208

[75] Whyte MP, McAlister WH, Patton LS, Magill HL, Fallon MD, Lorentz WB, et al. Enzyme replacement therapy for infantile hypophosphatasia attempted by intravenous infusions of alkaline phosphatase-rich Paget plasma: Results in three additional patients. The Journal of Pediatrics. 1984;**105**(6):926-933

[76] Whyte MP, Magill HL, Fallon MD, Herrod HG. Infantile hypophosphatasia:

Normalization of circulating bone alkaline phosphatase activity followed by skeletal remineralization. Evidence for an intact structural gene for tissue nonspecific alkaline phosphatase. The Journal of Pediatrics. 1986;**108**(1):82-88

[77] Weninger M, Stinson RA, Plenk H, Böck P, Pollak A. Biochemical and morphological effects of human hepatic alkaline phosphatase in a neonate with hypophosphatasia. Acta Paediatrica Scandinavica. Supplement. 1989;**360**:154-160

[78] Narisawa S, Fröhlander N, Millán JL. Inactivation of two mouse alkaline phosphatase genes and establishment of a model of infantile hypophosphatasia. Developmental Dynamics. 1997;**208**(3):432-446

[79] Millán JL, Narisawa S, Lemire I, Loisel TP, Boileau G, Leonard P, et al. Enzyme replacement therapy for murine hypophosphatasia. Journal of Bone and Mineral Research. 2008;**23**(6):777-787

[80] Whyte MP. Hypophosphatasia: Enzyme replacement therapy brings new opportunities and new challenges. Journal of Bone and Mineral Research. 2017;**32**(4):667-675

[81] Administration UFaD. Highlights of Prescribing Information. Available from: https://www.accessdata.fda.gov/drugsatfda_docs/label/2020/125513s018lbl.pdf

[82] STRENSIQ (Asfotase Alfa) Solution for Subcutaneus Injection [Internet]. 2015 [Cited 2024, May 4]. Available from: https://www.accessdata.fda.gov/drugsatfda_docs/nda/2015/125513Orig1s000TOC.cfm

[83] Brind'Amour K. What Pediatric Subspecialists Need to Know about Hypophosphatasia and its Treatment 2019 [Updated 2019, Sep 26; Cited 2024 June 12]. Available from: https://pediatricsnationwide.org/2019/09/26/what-pediatric-subspecialists-need-to-know-about-hypophosphatasia-and-its-treatment/

[84] Pharma A. Annex1 Summary of Product Characteristics 2017 [Available from: Strensiq [Summary of Product Characteristics]]. Rueil-Malmaison, France: Alexion Europe; June 8, 2017

[85] Taketani T, Oyama C, Mihara A, Tanabe Y, Abe M, Hirade T, et al. Ex vivo expanded allogeneic mesenchymal stem cells with bone marrow transplantation improved Osteogenesis in infants with severe Hypophosphatasia. Cell Transplantation. 2015;**24**(10):1931-1943

[86] Urnov FD, Miller JC, Lee YL, Beausejour CM, Rock JM, Augustus S, et al. Highly efficient endogenous human gene correction using designed zinc-finger nucleases. Nature. 2005;**435**(7042):646-651

[87] Porteus MH, Baltimore D. Chimeric nucleases stimulate gene targeting in human cells. Science. 2003;**300**(5620):763

[88] Bulcha JT, Wang Y, Ma H, Tai PWL, Gao G. Viral vector platforms within the gene therapy landscape. Signal Transduction and Targeted Therapy. 2021;**6**(1):53

[89] Arabi F, Mansouri V, Ahmadbeigi N. Gene therapy clinical trials, where do we go? An overview. Biomedicine & Pharmacotherapy. 2022;**153**:113324

[90] Porter DL, Levine BL, Kalos M, Bagg A, June CH. Chimeric antigen receptor-modified T cells in chronic lymphoid leukemia. The New England Journal of Medicine. 2011;**365**(8):725-733

[91] Fischer A, Hacein-Bey S, Le Deist F, Soudais C, Di Santo JP, de Saint BG, et al. Gene therapy of severe combined immunodeficiencies. Immunological Reviews. 2000;**178**:13-20

[92] Cavazzana-Calvo M, Hacein-Bey S, de Saint BG, Gross F, Yvon E, Nusbaum P, et al. Gene therapy of human severe combined immunodeficiency (SCID)-X1 disease. Science. 2000;**288**(5466):669-672

[93] Nienhuis AW, Dunbar CE, Sorrentino BP. Genotoxicity of retroviral integration in hematopoietic cells. Molecular Therapy. 2006;**13**(6): 1031-1049

[94] Raper SE, Chirmule N, Lee FS, Wivel NA, Bagg A, Gao GP, et al. Fatal systemic inflammatory response syndrome in a ornithine transcarbamylase deficient patient following adenoviral gene transfer. Molecular Genetics and Metabolism. 2003;**80**(1-2):148-158

[95] Wilson JM. Lessons learned from the gene therapy trial for ornithine transcarbamylase deficiency. Molecular Genetics and Metabolism. 2009;**96**(4):151-157

[96] Liu Z, Shi M, Ren Y, Xu H, Weng S, Ning W, et al. Recent advances and applications of CRISPR-Cas9 in cancer immunotherapy. Molecular Cancer. 2023;**22**(1):35

[97] High KA, Roncarolo MG. Gene therapy. The New England Journal of Medicine. 2019;**381**(5):455-464

[98] Parums DV. Editorial: First regulatory approvals for CRISPR-Cas9 therapeutic gene editing for sickle cell disease and transfusion-dependent beta-thalassemia. Medical Science Monitor. 2024;**30**:e944204

[99] Khan AN, Chowdhury A, Karulkar A, Jaiswal AK, Banik A, Asija S, et al. Immunogenicity of CAR-T cell therapeutics: Evidence, mechanism and mitigation. Frontiers in Immunology. 2022;**13**:886546

[100] Packiam VT, Lamm DL, Barocas DA, Trainer A, Fand B, Davis RL 3rd, et al. An open label, single-arm, phase II multicenter study of the safety and efficacy of CG0070 oncolytic vector regimen in patients with BCG-unresponsive non-muscle-invasive bladder cancer: Interim results. Urologic Oncology. 2018;**36**(10):440-447

[101] Administration UFaD. Luxturna 2022 [Updated 2022 Sep 06; Cited 2024 May 20]. Available from: https://www.fda.gov/vaccines-blood-biologics/cellular-gene-therapy-products/luxturna

[102] Agency EM. Upstaza. Available from: https://www.ema.europa.eu/en/medicines/human/EPAR/upstaza

[103] Verdera HC, Kuranda K, Mingozzi F. AAV vector immunogenicity in humans: A long journey to successful gene transfer. Molecular Therapy. 2020;**28**(3):723-746

[104] Horton RH, Saade D, Markati T, Harriss E, Bonnemann CG, Muntoni F, et al. A systematic review of adeno-associated virus gene therapies in neurology: The need for consistent safety monitoring of a promising treatment. Journal of Neurology, Neurosurgery, and Psychiatry. 2022;**93**(12):1276-1288

[105] Mendell JR, Al-Zaidy SA, Rodino-Klapac LR, Goodspeed K, Gray SJ, Kay CN, et al. Current clinical applications of In vivo gene therapy with AAVs. Molecular Therapy. 2021;**29**(2):464-488

[106] Shieh PB, Kuntz NL, Dowling JJ, Muller-Felber W, Bonnemann CG, Seferian AM, et al. Safety and efficacy of gene replacement therapy for X-linked myotubular myopathy (ASPIRO): A multinational, open-label, dose-escalation trial. Lancet Neurology. 2023;**22**(12):1125-1139

[107] Lek A, Wong B, Keeler A, Blackwood M, Ma K, Huang S, et al. Death after High-dose rAAV9 gene therapy in a patient with Duchenne's muscular dystrophy. The New England Journal of Medicine. 2023;**389**(13):1203-1210

[108] Salabarria SM, Corti M, Coleman KE, Wichman MB, Berthy JA, D'Souza P, et al. Thrombotic microangiopathy following systemic AAV administration is dependent on anti-capsid antibodies. The Journal of Clinical Investigation. 2024;**134**(1):e173510

[109] Boutin S, Monteilhet V, Veron P, Leborgne C, Benveniste O, Montus MF, et al. Prevalence of serum IgG and neutralizing factors against adeno-associated virus (AAV) types 1, 2, 5, 6, 8, and 9 in the healthy population: Implications for gene therapy using AAV vectors. Human Gene Therapy. 2010;**21**(6):704-712

[110] Watanabe S, Kanatsu-Shinohara M, Ogonuki N, Matoba S, Ogura A, Shinohara T. In vivo genetic manipulation of Spermatogonial stem cells and their microenvironment by adeno-associated viruses. Stem Cell Reports. 2018;**10**(5):1551-1564

[111] Honaramooz A, Megee S, Zeng W, Destrempes MM, Overton SA, Luo J, et al. Adeno-associated virus (AAV)-mediated transduction of male germ line stem cells results in transgene transmission after germ cell transplantation. The FASEB Journal. 2008;**22**(2):374-382

[112] Kohn DB, Booth C, Shaw KL, Xu-Bayford J, Garabedian E, Trevisan V, et al. Autologous ex vivo Lentiviral gene therapy for adenosine deaminase deficiency. The New England Journal of Medicine. 2021;**384**(21):2002-2013

[113] Milone MC, O'Doherty U. Clinical use of lentiviral vectors. Leukemia. 2018;**32**(7):1529-1541

[114] Senior M. After Glybera's withdrawal, what's next for gene therapy? Nature Biotechnology. 2017;**35**(6):491-492

[115] Brooks PJ, Ottinger EA, Portero D, Lomash RM, Alimardanov A, Terse P, et al. The platform vector gene therapies project: Increasing the efficiency of adeno-associated virus gene therapy clinical trial Startup. Human Gene Therapy. 2020;**31**(19-20):1034-1042

[116] Sciences NCfAT. Bespoke Gene Therapy Consortium. Available from: https://ncats.nih.gov/research/research-activities/BGTC.

[117] Matsumoto T, Miyake K, Yamamoto S, Orimo H, Miyake N, Odagaki Y, et al. Rescue of severe infantile hypophosphatasia mice by AAV-mediated sustained expression of soluble alkaline phosphatase. Human Gene Therapy. 2011;**22**(11):1355-1364

[118] Yamamoto S, Orimo H, Matsumoto T, Iijima O, Narisawa S, Maeda T, et al. Prolonged survival and phenotypic correction of Akp2(−/−) hypophosphatasia mice by lentiviral gene therapy. Journal of Bone and Mineral Research. 2011;**26**(1):135-142

[119] Sugano H, Matsumoto T, Miyake K, Watanabe A, Iijima O, Migita M, et al. Successful gene therapy in utero for lethal murine hypophosphatasia. Human Gene Therapy. 2012;**23**(4):399-406

[120] Iijima O, Miyake K, Watanabe A, Miyake N, Igarashi T, Kanokoda C, et al. Prevention of lethal murine Hypophosphatasia by neonatal ex vivo gene therapy using Lentivirally transduced bone marrow cells. Human Gene Therapy. 2015;**26**(12):801-812

[121] Matsumoto T, Miyake K, Miyake N, Iijima O, Adachi K, Narisawa S, et al. Treatment with bone maturation and average lifespan of HPP model mice by AAV8-mediated neonatal gene therapy via single muscle injection. Molecular Therapy - Methods & Clinical Development. 2021;**22**:330-337

[122] Wu Z, Asokan A, Samulski RJ. Adeno-associated virus serotypes: Vector toolkit for human gene therapy. Molecular Therapy. 2006;**14**(3):316-327

[123] Powell SK, Rivera-Soto R, Gray SJ. Viral expression cassette elements to enhance transgene target specificity and expression in gene therapy. Discovery Medicine. 2015;**19**(102):49-57

[124] Wang Z, Zhu T, Qiao C, Zhou L, Wang B, Zhang J, et al. Adeno-associated virus serotype 8 efficiently delivers genes to muscle and heart. Nature Biotechnology. 2005;**23**(3):321-328

[125] Kinoshita Y, Mohamed FF, Amadeu de Oliveira F, Narisawa S, Miyake K, Foster BL, et al. Gene therapy using adeno-associated virus serotype 8 encoding TNAP-D(10) improves the skeletal and Dentoalveolar phenotypes in Alpl(−/−) mice. Journal of Bone and Mineral Research. 2021;**36**(9):1835-1849

[126] Foster BL, Kuss P, Yadav MC, Kolli TN, Narisawa S, Lukashova L, et al. Conditional Alpl ablation Phenocopies dental defects of Hypophosphatasia. Journal of Dental Research. 2017;**96**(1):81-91

[127] Amadeu de Oliveira F, Mohamed FF, Kinoshita Y, Narisawa S, Farquharson C, Miyake K, et al. Gene therapy using recombinant AAV type 8 vector encoding TNAP-D(10) improves the skeletal phenotypes in murine models of Osteomalacia. JBMR Plus. 2023;**7**(1):e10709

[128] Mohamed FF, Chavez MB, Huggins S, Bertels J, Falck A, Suva LJ, et al. Dentoalveolar defects of Hypophosphatasia are recapitulated in a sheep Knock-In model. Journal of Bone and Mineral Research. 2022;**37**(10):2005-2017